VOLUME ONE HUNDRED AND NINE

SEMICONDUCTORS AND SEMIMETALS

Brillouin Scattering Part 1

SERIES EDITORS

VOLUME ONE HUNDRED AND NINE

SEMICONDUCTORS AND SEMIMETALS

Brillouin Scattering Part 1

Edited by

BENJAMIN J. EGGLETON
School of Physics,
University of Sydney Nano Institute (Sydney Nano),
University of Sydney, NSW, Australia

MICHAEL J. STEEL
School of Mathematical and Physical Sciences,
Macquarie University, Sydney, NSW, Australia

CHRISTOPHER G. POULTON
School of Mathematical and Physical Sciences,
University of Technology Sydney, Sydney, NSW, Australia

ELSEVIER

ACADEMIC PRESS
An imprint of Elsevier

Academic Press is an imprint of Elsevier
50 Hampshire Street, 5th Floor, Cambridge, MA 02139, United States
525 B Street, Suite 1650, San Diego, CA 92101, United States
The Boulevard, Langford Lane, Kidlington, Oxford OX5 1GB, United Kingdom
125 London Wall, London, EC2Y 5AS, United Kingdom

First edition 2022

ISBN: 978-0-323-98929-9
ISSN: 0080-8784

For information on all Academic Press publications
visit our website at https://www.elsevier.com/books-and-journals

Publisher: Zoe Kruze
Acquisitions Editor: Jason Mitchell
Developmental Editor: Jhon Michael Peñano
Production Project Manager: Abdulla Sait
Cover Designer: Greg Harris

Typeset by STRAIVE, India

Contents

Contributors for Volume 1

Christopher G. Baker
Australian Research Council Centre of Excellence for Engineered Quantum Systems, School of Mathematics and Physics, The University of Queensland, St Lucia, QLD, Australia

Jean-Charles Beugnot
FEMTO-ST Institute, UMR 6174 CNRS, Universite Bourgogne Franche Comte, Besançon, France

Warwick P. Bowen
Australian Research Council Centre of Excellence for Engineered Quantum Systems, School of Mathematics and Physics, The University of Queensland, St Lucia, QLD, Australia

Paulo Dainese
Corning Research and Development Corporation, Corning, NY, United States

Jack G.E. Harris
Department of Physics, Yale University, New Haven, CT, United States

Benjamin J. Eggleton
School of Physics, University of Sydney Nano Institute (Sydney Nano), University of Sydney, NSW, Australia

Raphaël Van Laer
Department of Microtechnology and Nanoscience (MC2), Chalmers University of Technology, Gothenburg, Sweden

Stephen J. Madden
Laser Physics Center, Australian National University, Canberra, ACT, Australia

Meng Pang
Shanghai Institute of Optics and Fine Mechanics, Chinese Academy of Sciences, Shanghai, China

Ravi Pant
School of Physics, Indian Institute of Science Education and Research (IISER), Thiruvananthapuram, Kerala, India

Christopher G. Poulton
School of Mathematical and Physical Sciences, University of Technology Sydney, Sydney, NSW, Australia

Mikołaj K. Schmidt
Macquarie University Research Centre in Quantum Engineering (MQCQE), MQ Photonics Research Centre, School of Mathematical and Physical Sciences, Macquarie University, Sydney, NSW, Australia

Heedeuk Shin
Department of Physics, Pohang University of Science and Technology (POSTECH), Pohang, South Korea

Michael J. Steel
School of Mathematical and Physical Sciences, Macquarie University, Sydney, NSW, Australia

Luc Thévenaz
Gleb Wataghin Physics Institute, University of Campinas, Campinas, SP, Brazil

Gustavo Wiederhecker
Group for Fibre Optics, Ecole Polytechnique Fédérale de Lausanne (EPFL), Lausanne, Switzerland; Gleb Wataghin Physics Institute, University of Campinas, Campinas, SP, Brazil

Christian Wolff
Center for Nano Optics, University of Southern Denmark, Odense M, Denmark

Fan Yang
Ecole Polytechnique Fédérale de Lausanne (EPFL), Lausanne, Switzerland

Preface

The year of this book's release marks the centenary of Léon Brillouin's seminal 1922 paper (Brillouin, 1922), which for the first time provided a detailed theory of the effect that now bears his name. As one might expect, the history of the prediction (and subsequent observation) of Brillouin scattering is more nuanced than the publication of a single paper, and the contemporary view is that both Brillouin and Leonid Mandelstam played vital roles in the formulation of Brillouin scattering theory. One hundred years later, our understanding of this effect plays a critical role in many areas of modern science and technology, particularly lasers, communications, fiber optics, sensing, and imaging applications as well as in experimental physics.

The history of Brillouin scattering is rich and multifaceted and is linked to other significant developments in optical science, such as the invention of the laser, fiber optics, and the rise of integrated photonics. Brillouin scattering, first observed in the 1930s, with early demonstrations focused on characterizing liquids and gases, became an accessible and powerful tool following the invention of the laser, which enabled the transition to stimulated Brillouin scattering (SBS). Following the invention of optical fibers, the detrimental effect of SBS on transport of narrow linewidth signals was quickly recognized, motivating significant research efforts focused on suppressing SBS. During this period, many underlying principles for manipulating SBS in waveguides were developed, offering new insights for enhancement and application of SBS. For example, it was realized that SBS can form the basis of a novel form of optical gain, without the requirement for "lasing materials," an effect now simply referred to as Brillouin lasing. SBS lasers and oscillators have been extensively studied for decades in the form of optical fibers, and more recently in integrated waveguides and resonators. SBS can also form the basis of sensor modalities, particularly when incorporated into optical fibers, which can extend over long distances and be integrated into infrastructure; SBS-based distributed sensors have been extensively deployed and are the most successful commercially of all the SBS applications.

The SBS research community has recently seen a renaissance in basic research and applications, enabled by advances in nanotechnology, particularly through the developments of complex photonic crystal fibers (PCFs) and photonic integrated circuits, and the simultaneous development of new nonlinear materials, such as chalcogenide glass, silicon, and, very recently, thin-film lithium niobate. This new wave of activity is opening possibilities for observing and harnessing SBS on length scales of centimeters and even millimeters, which is compatible with photonic integrated circuits. Furthermore, an increasing number of materials platforms have been developed that allow for strong confinement of both optical waves and acoustic waves, providing completely new degrees of freedom to exploit the interaction of optical waves and acoustic waves. In these new geometries, there is a fascinating convergence of SBS with optomechanics—the nonlinear interaction between light and stationary discrete acoustic modes. The theoretical descriptions of these seemingly disconnected physical phenomena have strong parallels, suggesting new directions for investigations in both fields.

On the more technological side, there has been a paradigm shift in photonic signal processing based on using SBS effects for manipulating both digital and analog information signals with unprecedented resolution and bandwidth. The narrow linewidth Brillouin gain or loss that is usually a nuisance in lasers thus becomes the basis of a very powerful signal processing filter with unparalleled frequency resolution. When combined with the possibility of integrating SBS on to photonic chips, this has prompted developments translatable into applications in radar systems, communication systems, and next-generation 5G.

The significance of this publication anniversary in 2022, combined with the current research renaissance, motivates this compilation of chapters bringing together the world's leading scholars in the area to outline our understanding of the fundamental and theoretical aspects of SBS and explore the scientific and technological applications and implications of this knowledge. From its first conception, it has been our intention that this volume should present a broad perspective on the key thematic areas, and thus we have asked at least three authors from different institutions to collaborate on each chapter. Our hope is that in this way the book will present a balanced and considered perspective on the state of the art, how the field has evolved, and where it is headed.

We have been delighted by the enthusiasm of our contributors for this unusual approach to assembling a contributed book, and hope that it may become more common.

Finally, with a mix of historical, theoretical, and applications-oriented chapters, we believe this book will play a significant pedagogical role in bringing new scientists and engineers into the field, whether at the beginning of their research careers, or established researchers seeking to explore a new area.

With many contributions, we have organized the chapters into two volumes: (i) Volume 1, emphasizing more fundamental aspects; and (ii) Volume 2, emphasizing applications and technological considerations.

In Chapter 1, Poulton, Steel, Wiederhecker, and Thévenaz introduce this first volume with a broad overview of Brillouin scattering, combining a historical perspective with a largely nontechnical overview of the underlying physics of SBS. With this background, we hope to give a broad context to the remainder of the volume. The historical perspective traces the history of Brillouin scattering from the second decade of the 20th century when both Brillouin and Mandelstam independently formulated theories of optical scattering by high-frequency acoustic waves. The chapter proceeds to review early efforts to observe Brillouin scattering, highlighting some of the technical challenges in the very early experiments, where Raman scattering was dominant. With the invention of the laser in the 1960s, SBS was measured and shortly thereafter was seen in optical fibers, which led to a vibrant research community that was prominent at photonics conferences for several decades. The chapter then briefly surveys the more recent period during which SBS has been extensively studied and harnessed in PCFs and integrated waveguides. The second part of the chapter describes, at a high level and largely qualitatively, how Brillouin scattering is now understood physically in addition to providing an overview of the categories of Brillouin scattering that may be observed in modern experiments.

In Chapter 2, Wolff, Poulton, Steel, and Wiederhecker present a comprehensive theoretical formalism for SBS. This includes a detailed derivation of the coupled mode equations for Brillouin scattering in the popular Hamiltonian formalism, a detailed analysis of the different kinds of behavior in the various classes of Brillouin scattering, a parallel formalism for optoelastic scattering in resonators, and an explanation of the connection of the modern formulation to the well-known hydrodynamic treatment based on pressure fields that has dominated the fiber literature.

In Chapter 3, Schmidt, Baker, and Van Laer explore how the recent renaissance in Brillouin scattering research that emerged in the context of subwavelength scale waveguides has brought about a convergence of two separate communities studying interactions between light and sound. In the community of *cavity optomechanics*, scientists are mostly concerned with interactions between localized light and sound in optically resonant and few-moded devices. In the community of *Brillouin scattering*, the focus is on characterizing and harnessing interactions between traveling light and sound in optically broadband and many-moded structures. Nevertheless, there are many parallels in the physics and formalism, and this chapter explores them in detail.

In Chapter 4, Dainese, Beugnot, Yang, and Pang discuss the properties of SBS in PCFs in which simultaneous guidance of both optical and mechanical waves is provided by the tight transverse confinement and relatively long propagation distances. Both solid- and hollow-core PCFs have been demonstrated as platforms for different aspects of Brillouin scattering, and numerous applications are being explored.

In Chapter 5, Pant, Madden, Eggleton, and Shin review the development of integrated planar waveguides in which SBS can be induced and harnessed on centimeter scales. They then discuss the requirements for realizing and fabricating SBS in chip-scale formats and highlight emerging applications that have been demonstrated, including slow-light, frequency comb generation, and microwave photonic filters. Finally, they provide a brief overview of hybrid integration with CMOS-compatible silicon integration.

In Chapter 6, Bowen, Baker, and Harris review recent research on Brillouin scattering in quantum fluids, specifically superfluid helium, which reveals quantum behaviors that are not present in regular fluids. The most notable is the vanishing viscosity of superfluid helium at low temperatures, which, when combined with other extreme material properties, provides a regime to harness SBS for new quantum technologies, such as sensors and memories.

In Chapter 7, which begins Volume 2, Zadok, Bao, Yang, and Thevenaz review the principles of harnessing SBS in waveguides for sensors, including point sensors and distributed sensors. The chapter highlights the many successful commercial applications of Brillouin scattering with examples of Brillouin fiber sensors that span hundreds of kilometers, obtain submillimeter resolution, identify subdegree centigrade temperature changes and tens of parts per million strain events, and track dynamic signals at kilohertz and even megahertz rates.

In Chapter 8, Merklein, Schneider, and Vahala review Brillouin-based radiofrequency sources. This approach relies on the use of Brillouin scattering to generate multitone optical signals by creating an optical Stokes wave that is separated from the optical pump by around 10 GHz and can then be used to generate a stable RF beat signal. The chapter introduces key performance metrics such as phase noise, followed by a perspective on current trends and future directions.

In Chapter 9, Marpaung and Liu discuss applications of SBS to microwave photonics. They describe how SBS has been harnessed for microwave photonic signal processing due to its distinct advantages in bridging between high-resolution and coarse signal processing. These are not possible with other competing technologies; the MHz-level linewidth enabled by SBS allows for high-resolution filtering tunable over ultrawideband optical frequencies, unmatched by most on-chip devices.

In Chapter 10, Blumenthal, Kabakova, Rakich, and Vahala review recent research on integrated and micro-Brillouin lasers to bring the robustness, light weight and small size, and low cost associated with integration, to a wide range of applications that require the optical spectral purity and signal processing capabilities of Brillouin light-sound interactions.

In Chapter 11, Corcoran and Choudhary review SBS in optical communication systems, highlighting the advantages and challenges of applying SBS in this context. They discuss how SBS has been explored as a tool to enable novel functions in optical communication systems, but also limits power transmission, which in turn limit achievable signal-to-noise ratio—the key metric for increasing data rates.

In Chapter 12, Stiller, Winful, Boyd, and Merklein review principles of slow light, dynamic gratings, and light storage based on SBS. They discuss the need to delay or buffer optical signals in high-speed optical data transmission systems to permit the signal processing, switching, and rerouting that take place in the nodes of optical networks. The buffering is ideally an all-optical process as it avoids electro-optical conversion and maintains a large bandwidth.

In Chapter 13, Otterstrom, Kittlaus, Sohn, and Bahl discuss nonreciprocity in Brillouin scattering, including the underlying physical and mathematical principles and demonstrations of nonreciprocal functions, which can be the basis of isolators and circulators, key components in photonic systems. Such devices, which are key components in photonic integrated circuits, are traditionally implemented using magneto-optic components that are not always compatible with photonic integration techniques.

In Chapter 14, Li, Florez, Pan, Madiot, Sotomayor Torres, and Li discuss electromechanical Brillouin scattering (EBS), which deals with Brillouin scattering of light by electromechanically excited elastic waves. They explain that the main difference between EBS and SBS is that in EBS the elastic waves (i.e., sound or phonons) are generated by an electromechanical transducer using piezoelectric materials, which convert electromagnetic waves to propagating elastic waves or, in some cases, localized mechanical modes.

Finally, in Chapter 15, Kabakova, Scarcelli, and Yun describe applying Brillouin light scattering (BLS) to measure the mechanical properties of tissues and cells to help understand the roles of biomechanics for advancing life sciences and medical diagnosis and treatments. They show that unlike traditional mechanical testing techniques, BLS does not require application of extrinsic forces nor the use of artificially introduced labels, making this technology particularly suitable for *in situ* and *in vivo* settings.

This hugely diverse set of contributions emphasizes the ever-growing impact that Brillouin-Mandelstam scattering is delivering, a century after its first conception. We trust that readers will be excited by the range of activity and applications and inspired in driving the field forward. Finally, we sincerely thank all the contributors for their efforts in preparing a wide range of insightful and erudite chapters and for doing so with such enthusiasm during this most challenging couple of years. We strongly believe the result is a fitting vehicle with which to say Happy 100th, Brillouin-Mandelstam scattering.

Benjamin J. Eggleton
School of Physics, University of Sydney Nano Institute (Sydney Nano), University of Sydney, NSW, Australia

Michael J. Steel
School of Mathematical and Physical Sciences, Macquarie University, NSW, Australia

Christopher G. Poulton
School of Mathematical and Physical Sciences, University of Technology Sydney, NSW, Australia

List of symbols

The following table lists a number of frequently occurring symbols in this book.

Symbol	Notes	Description	Unit
α_n, α_b	(a)	Power loss constants of optical and elastic modes along the waveguide	1/m
γ_e		Electrostrictive constant	1
Γ_n, Γ_b	(b)	Amplitude decay constants of optical and elastic modes in time	1/s
ε_{ij}	(c,d)	Relative dielectric permittivity tensor	1
η_{ijlm}		Dynamic viscosity (Kelvin-Voigt) tensor	$\mathrm{Ns/m^2}$
$\boldsymbol{\pi}$		Local momentum density	$\mathrm{Ns/m^3}$
ρ		(Local) mass density	$\mathrm{kg/m^3}$
σ_{ij}		Electric (Ohmic) conductance tensor°	m/Ω
$\lvert\Phi\rangle$		Elastic state	–
$\chi_{ij}, \chi^{(2)}_{ijk}, \chi^{(3)}_{ijkl}$		Linear, Pockels, and Kerr susceptibilities	1, m/V, $\mathrm{m^2/V^2}$
$\lvert\Psi_n\rangle$		Optical states	–
ω_n		Optical (angular) frequencies	1/s
Ω		Elastic (angular) frequency	1/s
$\hat{a}_n$, $\hat{a}^\dagger_n$, a_n	(e,f)	Ladder operators and classical amplitude of optical mode	Depends on argument
$\hat{b}$, $\hat{b}^\dagger$, b	(e,f)	Ladder operators and classical amplitude of elastic mode	Depends on argument
$\boldsymbol{B}$, $\boldsymbol{D}$, $\boldsymbol{E}$, $\boldsymbol{H}$	(g)	Real electromagnetic fields and inductions	T, $\mathrm{C/m^2}$, V/m, A/m
$\boldsymbol{b}$, $\boldsymbol{d}$, $\boldsymbol{e}$, $\boldsymbol{h}$	(g)	Complex electromagnetic fields and inductions	$\mathrm{C/m^2}$, V/m, A/m

Continued

—cont'd

Symbol	Notes	Description	Unit
c_n, c_{b}; c_{vac}		Phase velocity or optical and elastic modes; vacuum speed of light	m/s
c_{ijlm}		Mechanical stiffness tensor	N/m^2
$\mathcal{E}_n$, $\mathcal{E}_{\text{b}}$	(f)	Optical and elastic mode normalization energies	J
f		Complex force density	N/m^3
$\boldsymbol{F}$		Lumped force	N
g	(a)	Brillouin power gain density of a material	m/W
G	(a)	Modal Brillouin power gain	1/(Wm)
i, j, k, l		Indices that can assume the Cartesian coordinates $\{x, y, z\}$	–
n	(h)	Label for numbering optical modes (band index)	–
k_n		Optical wavenumbers	1/m
L		Quantization length for QM waveguide theory	m
p_{ijkl}		Photoelastic Pockels tensor	1
$\boldsymbol{P}$	(f)	Electric polarisation (mostly nonlinear)	C/m^2
$\mathcal{P}_n$, $\mathcal{P}_{\text{b}}$		Power flux of optical/elastic modes through x-y-plane	W
q		Elastic wavenumber	1/m
Q		Opto-mechanical coupling coefficient	J
$\boldsymbol{r}$		Position vector $\boldsymbol{r} = (x, y, z)^T$	m
$\overleftrightarrow{R}$; R_{ij}		Rotation matrix, also local rotation field	1
s_{ijkl}		Mechanical compliance tensor	m^2/N
$\overleftrightarrow{S}$; S_{ij}	(c)	Mechanical (symmetric) strain tensor	1
$\mathcal{S}_n$, $\mathcal{S}_{\text{b}}$		Poynting flux of optical and elastic modes	W/m^2
t		Time	s
$\overleftrightarrow{T}$; T_{ij}	(c)	Mechanical stress tensor	N/m^2

—cont'd

Symbol	Notes	Description	Unit
$\boldsymbol{u}$	(g)	Complex mechanical displacement field	m
$\boldsymbol{U}$	(g)	Real mechanical displacement field	m
v_n, v_b		Optical and elastic group velocities	m/s
$\boldsymbol{V}$		Mechanical velocity field	m/s
x, y, z	(e)	The Cartesian coordinates	m

Notes:

(a) Divide by two when used in amplitude equation.

(b) Multiply by two when used in power equation; divide by 2π for respective amplitude line width.

(c) Avoid using σ and ϵ for stress and strain, respectively.

(d) To avoid confusion: ε_{ij} is the relative dielectric tensor, explicitly include ϵ_0 when required.

(e) By convention the main direction of propagation should be along the positive z-axis.

(f) Replacing a_n, b with $\hat{a}_n, \hat{b}$ and setting $\mathcal{E}_n = \hbar\omega_n/L$, $\mathcal{E}_b = \hbar\Omega/L$ should yield QM formulation.

(g) We tend to use upper case for real time-domain fields, lower case for complex frequency-domain fields.
Positive frequency parts go as $e^{-i\omega t}$. So the real-valued displacement field is $\boldsymbol{U}(\boldsymbol{r}, t) = \boldsymbol{u}(\boldsymbol{r})e^{-i\omega t} + \boldsymbol{u}^*(\boldsymbol{r})e^{i\omega t} = \boldsymbol{u}(\boldsymbol{r})e^{-i\omega t} + \text{c.c.}$

(h) Alternatively use subscripts (S) or (P) to label an optical pump and AS/S for (anti-)Stokes.

CHAPTER ONE

Historical perspective and basic principles

Christopher G. Poulton[a,*], Michael J. Steel[b], Gustavo Wiederhecker[d], and Luc Thévenaz[c]

[a]School of Mathematical and Physical Sciences, University of Technology Sydney, Sydney, NSW, Australia
[b]School of Mathematical and Physical Sciences, Macquarie University, Sydney, NSW, Australia
[c]Gleb Wataghin Physics Institute, University of Campinas, Campinas, SP, Brazil
[d]Group for Fibre Optics, Ecole Polytechnique Fédérale de Lausanne (EPFL), Lausanne, Switzerland
*Corresponding author: e-mail address: christopher.poulton@uts.edu.au

Contents

1. Introduction

The interaction of light and matter has been a topic of fascination for skygazers and philosophers for thousands of years. The influence of *matter on light* has been considered since ancient times: geometric optics dates from the

Semiconductors and Semimetals, Volume 109
ISSN 0080-8784
https://doi.org/10.1016/bs.semsem.2022.04.001

3rd to 4th century BCE in the works of Euclid (see Burton, 1945) and Mozi (see Needham, 1974). Euclid's writings led, via Ptolemy's *Optics* (see Smith, 1996), to the work of Ibn Sahl (see Rashed, 1990), who appears to have been the first to quantitatively enunciate the laws of refraction. However, consideration of the possibility of a converse interaction, that of *light on matter*, only appeared much later. In 1619, Kepler proposed that the orientation of the tails of comets away from the sun was the result of light pressure exerted by the sun, and although Newton was not convinced by this idea, Leonard Euler in 1746 showed that a longitudinal wave would indeed produce a pressure in the observed direction. Light is of course a transverse oscillation, but in 1873, Maxwell used his new theory of electromagnetism to show that light waves should carry momentum and would exert a pressure on material surfaces given by $p = I/c$, for irradiance I and speed of light c. This prediction was ultimately confirmed by the elegant experiments of Lebedev (1901) and Nichols and Hull (1903).[a]

In the years spanning the First World War, a new question was posed independently by the physicists Léon Brillouin and Leonid Mandelstam: What is the effect on light when the material part of the light–matter interaction takes the form of an acoustic wave? In tackling this question, these scientists shed new light on the nature of solids in the period between the old and modern quantum theories, and in particular, provided a pathway to validate P. Debye's theory of elastic vibrations and the specific heat of solids (Debye, 1912). These two investigations, published in 1922 and 1926 (Brillouin, 1922; Mandelstam, 1926) (following an earlier less-developed study by Brillouin (1914)) resulted in major new insights that have now driven 100 years of research.

This chapter aims to provide the reader with an appreciation of the historical context of Brillouin scattering, from its origins and early experiments to the present day, as well as a high-level overview of exactly what we mean by Brillouin–Mandelstam scattering, and how it arises. To this end we first cover the context of Brillouin and Mandelstam's research, together with a summary of the main findings of their papers and a discussion of the early experiments. We then give a highly abbreviated history of the evolution of Brillouin scattering since 1922 to the present day. While we have done our best to present an accurate account, any history, especially one written by physicists rather than historians, is necessarily incomplete and colored by the personal perspectives and interests of its authors. We therefore urge

[a] Readers interested in the history of radiation pressure are directed to the introduction of the paper by Nichols and Hull (1903), which gives an insightful overview of earlier experiments.

interested readers to sample some of the first- and second-hand accounts we have drawn on, including those by Thomas (1985), Rank (1971), Singh and Riess (2001), and Feinberg (2002), a complete biography of Mandelstam by Pechenkin (2019), E. Garmire's commentary on the early years of stimulated Brillouin scattering (SBS; Garmire, 2017), and reflections by Brillouin himself (Brillouin, 1969). Close study of the original Brillouin and Mandelstam papers is also very rewarding. The authors are happy to forward translations of these two works to interested readers.

In the second part of the chapter, we turn to the present and describe a modern overview of the physical principles of Brillouin scattering at the level of a phenomenological description of the process, as well as a survey of the key configurations that are encountered.

2. Origins and discovery of Brillouin–Mandelstam scattering

The year of this book's release marks the centenary of Léon Brillouin's seminal 1922 paper (Brillouin, 1922) which provided a detailed theory of the effect that now bears his name. In fact as we discuss below, some key elements of the scattering process were described in Brillouin's very first paper in 1914 (Brillouin, 1914), however, as he recounted late in life (Brillouin, 1969), a clear physical interpretation of these results was lacking in this earlier work.

It is essential that we also describe and appropriately acknowledge Leonid Mandelstam's independent insight that there should be a measurable interaction between light and thermally excited high frequency sound waves. That Mandelstam had framed the concept and mechanism for this interaction in 1918 (Feinberg, 2002), 4 years before Brillouin's second and decisive paper on the topic,[b] is not controversial. However, difficult living conditions in revolutionary Russia (and a conservative approach to publication that characterized his mature career) meant that Mandelstam did not publish this work until 1926 (Mandelstam, 1926). In parts of the literature, particularly from Russia and Eastern Europe, both the terms Mandelstam–Brillouin and Brillouin–Mandelstam scattering remain in wide use. Following this historical discussion, the remainder of this chapter (and the two volumes of this collection as a whole,) will adopt the standard and shorter term "Brillouin scattering."

[b] For the record, we note that Brillouin's paper was submitted in February 1921.

In this section, we provide a highly abridged picture of the approach of these two great 20th century scientists to the problem of opto-acoustic scattering, and some context for their interest in the topic within their larger careers.

2.1 A tale of two scientists

Born in 1879 in Mogilev in modern Belarus, Leonid Isaakovich Mandelstam began his university studies in Odessa, but his political activity as a student soon led to a move to Strasbourg in 1899. There he completed his physics training and enjoyed 14 successful years as a young theorist rising to full professor. Much of his work in this period was on prominent problems in classical optics (apparently including the first prediction of the phenomenon of frustrated total internal reflection (Feinberg, 2002)).

In the decades after the appearance of Maxwell's theory of electromagnetism, and through the "old-quantum" period of the first two decades of the 20th century, the nature of light–matter scattering was very much in vogue, with the origin of dispersion, absorption, and scattering all open questions. Today, undergraduates learn mostly of Rayleigh's famous treatment of the wavelength dependence of scattering by molecules (see Rayleigh (1899) and many earlier works), but the subject turned out to be rather more subtle, with the density fluctuations of the medium playing a key role, as suggested by Mandelstam in 1908 (Mandelstam, 1908). This was an area in which decisive contributions were subsequently made by Smoluchowski (1908) and Einstein (1910). While working in this area in the early part of his scientific career, Mandelstam became engaged in a number of intemperate and perhaps ill-advised debates with Planck and others, an experience that likely contributed to a cautious and patient approach to publication in the mature part of his career from the mid-1920s. He returned to Russia just as the First World War was breaking out, and the traumatic decade of revolution that followed led to a succession of unstable roles and very limited opportunity for publication.

Meanwhile, Léon Nicolas Brillouin, born in Sèvres, France in 1889, had been training in the top schools in Paris, enjoying lectures from Poincaré, Lorentz, Langevin, and Einstein (Thomas, 1985), and progressing to research in a range of topics including Brownian motion, properties of liquids and scattering, and the dispersion and diffraction of light. According to Brillouin's own recollections (Brillouin, 1969), his interest in scattering problems arose from considering the bright blue of the sky of Monte Rosa in the Alps; the problem of atmospheric scattering had recently been made famous by Rayleigh and its solution had stimulated much discussion, as we have already seen.

The original notion of scattering from vibrations in solids appears to have originated some years prior to Brillouin's 1922 paper. As he recalled himself (Brillouin, 1969), the idea arose from considering the recent scattering work from Rayleigh and Einstein, and Brillouin thought to apply this to solid bodies using Debye's recent (and very successful) theory of the specific heat of solids. The results of his calculations, in which Brillouin computed the scattering from longitudinal vibrations in a homogeneous body, were published in a note to the Comptes Rendus de L'Académie des Sciences of Paris in 1914 (Brillouin, 1914). In Eq. (6) of this paper there is a term that depends on the speed of sound, the wavelength of light, and the cosine of the angle between the incident ray of light and the scattered ray. Brillouin stated later that he spent some time puzzling over this term but was unable to make any progress on its physical meaning. Two months after the publication of this note, Brillouin's career was interrupted by the First World War, which he spent as a radio engineer in the French signal corps. On returning from the war in 1919 he found his notes (which he had tossed in a corner) in disarray: "...I was horrified when I tried to read my old notes. The computations were there, with no explanations—and they did not make sense to me any more. I had to rebuild everything from scratch." (Brillouin, 1969). Nevertheless he managed to answer his remaining questions and the theory was published in full in 1922 as the landmark paper (Brillouin, 1922) that truly launched the field.

Apparently, this new theory was not initially well received. According to Brillouin (1969):

> *The change in frequency was at first doubted or even denied: it had always been taken for granted that scattering did not involve any frequency change: "when there is a frequency change it is a problem of fluorescence, not scattering!" said my professors! Remember that Raman scattering was yet unknown, and appeared only in 1928!! The frequency change in Brillouin scattering was so small that it could be observed only with the best Perot-Fabry equipment.*

In comments reported by D.H. Rank (see Rank, 1971), Brillouin went further, saying that on presenting his theory in Strasbourg "he [Brillouin] was informed in no uncertain terms that his theory was sheer madness." To possibly account for some of this scepticism, it is worth noting that with still 3 or

4 years to pass before the beginnings of modern quantum theory and 6 years before Bloch's theorem on waves in periodic media, the physical reality of Debye's oscillators was not at all an established fact.

2.2 The Mandelstam and Brillouin papers

Let us now turn to the specific contributions of the two papers and make some remarks about their similarities and differences.

Both Brillouin's 1922 and Mandelstam's 1926 papers are fully classical calculations in the language of wave interference. However, the quantum concept of energy exchange between light and material oscillators that Planck and Einstein introduced two decades before is quite apparent. The strongest influence on both papers, aside from the light scattering literature mentioned above, must be the theories of specific heat of Einstein (1907) and particularly Debye (1912). Both papers follow Debye's insight that an optical medium supports a thermally excited ensemble of high frequency elastic waves. These waves will induce a time-variation in the local density and therefore also in the refractive index. Both papers develop far-field scattering problems to determine how an incoming optical plane wave will be modified by this time-varying index.

The principal prediction of both papers (we use Mandelstam's notation which is more transparent and closer to modern style) is that when an incoming plane light wave of frequency ν encounters a region of refractive index n supporting sound waves of speed a, the scattered light will include a doublet of shifted frequency components according to

$$\frac{\delta\nu}{\nu} = \pm 2\frac{a}{c}\sin\frac{\theta}{2}, \tag{1}$$

where $\delta\nu$ is the frequency shift, $c = c_{\text{vac}}/n$ is the speed of light in the medium, and θ is the angle between the direction of the incident beam and the observation point. Notably, the direction of travel of the sound wave does not enter explicitly (we revisit this in a moment with reference to Brillouin's paper). As Mandelstam notes[c]

> *When a plane light wave is diffracted due to a perturbation spreading in an unlimited elastic environment, the intensity of the light scattered in any particular direction changes periodically...The law of change does not depend on the form of the disturbance, but depends exclusively on the ratio of the speed of sound to the speed of light, the light period, and the angle formed by the observation direction with the direction of the incident wave (Mandelstam, 1926)*

[c] Google translation.

Mandelstam observes that while the expected shift in air is only $|\delta\nu/\nu| \approx 1.5 \times 10^{-6}$, the value for quartz is $|\delta\nu/\nu| \approx 4.5 \times 10^{-5}$, *"sufficient that the light intensity can be measured quite conveniently,"* though as we will see, the first observation was not achieved for another 4 years in 1930, over a decade after his original insights.

The details of the approaches in the two papers are quite different. Mandelstam's paper (much the shorter) treats both fields as scalar waves and is essentially kinematic, not attempting to calculate the overall scattering efficiency, nor any power dependence of the scattering with angle. Brillouin's analysis is more detailed in several respects. First, the light scattering is treated as a vector problem in Maxwell's theory (using a scattering formalism due to Einstein). In addition, the thermal excitation of the elastic field as a function of the elastic wave frequency is treated carefully using the Planck distribution and density of states factors obtained by Debye in his specific heat theory. Both the solid case (with transverse and longitudinal excitations) and the liquid case (with only longitudinal excitations) are considered. In this way, Brillouin is able to find the scattered power as a function of angle θ in terms of the photoelastic strength $\partial k/\partial \rho$, where k is the dielectric constant (his notation, which we follow here before switching below to the standard use of k as the optical wave number) and ρ the density.

Moreover, before his detailed calculation, Brillouin uses a constructive interference argument applied to scattering by a *stationary* density variation to note that the *scattering is only efficient* when the angle θ between the incoming and scattered ray satisfies[d]

$$2\Lambda \sin\frac{\theta}{2} = \frac{\lambda_0}{\sqrt{k}}, \tag{2}$$

or as we might say today

$$\sin\frac{\theta}{2} = \frac{1}{2}\frac{\Omega}{\omega}\frac{(c_{\text{vac}}/n)}{v_s}, \tag{3}$$

where v_s is the sound speed, while Ω and ω are the angular frequencies of the sound and light waves. Here is the expected phase-matching condition,

[d] Brillouin actually writes this angle as 2θ. We have adjusted the definition of this angle in order to be consistent with the choice of Mandelstam.

which will appear frequently in subsequent chapters. As Brillouin notes, this is simply a variant of the Bragg law.

Brillouin's remarks on this result are highly insightful:

> *This formula is very interesting, because it connects elastic and luminous waves whose wavelengths are the same order of magnitude. In particular, if we assume, as we will do further on, the medium is crossed by an infinity of elastic waves spreading in all directions, with all frequencies, the medium will diffuse light in all directions, but the light scattered in a direction that makes an angle θ with the incident ray will come uniquely from an elastic wave whose direction of propagation bisects the angle θ of the two rays, and whose wavelength Λ satisfies [the above] formula.*

He also turns the problem inside out in this remarkably prescient suggestion of using light as a probe of the elastic spectrum:

> *The measurement of the intensity of light diffused in all directions will therefore make it possible to know the intensity of sound waves of given directions and frequencies, and can be an extremely valuable way to analyze the thermal agitation of the elastic medium and measure the distribution of energy, at a given temperature, between the different elastic wave frequencies.*

One gets the strong sense that Brillouin saw this new scattering process as a means to directly confirm the presence of Debye's elastic waves.

2.3 First observations of Brillouin scattering

The small frequency shifts (of order 1 GHz) associated with Brillouin scattering made detection of the effect at the time extremely challenging. Mandelstam and his colleague Grigory Landsberg worked on the problem for several years, ultimately making the first observation of the related process of "combinational scattering" of light by *molecular* vibrations in fluids in February 1928, approximately 1 week before the same effect was observed by C.V. Raman. In a similar fashion to the long delay in publication of his 1918 Brillouin scattering ideas until 1926, Landsberg and Mandelstam (1928) held back publication of their combinational scattering work until May 1928 so as to make further observations and refine their theory (Feinberg, 2002). By this time Raman had published several well-cited papers (Raman, 1928; Raman and Krishnan, 1928a, b) and the community had firmly attached his name to the effect. This sequence of events likely cost Mandelstam a share in the 1930 Nobel Prize in Physics (Feinberg, 2002; Singh and Riess, 2001).

Because the frequency shift due to Raman scattering is several orders of magnitude larger than the Brillouin shift, it is not surprising that Raman scattering was observed first. The detection of Brillouin shifts of order 0.4 Å was extremely challenging for the spectrometers of the time. The first confirmed observations of the actual Brillouin–Mandelstam scattered doublet were ultimately reported by Evgeni Gross for half a dozen common fluids in a series of papers in 1930 (Gross, 1930a,b,c).[e] In fact, Gross, like several other researchers, was actually looking for very small-shift Raman-like effects due to molecular rotations rather than molecular vibrations, but soon realized he was observing shifts due to scattering of Debye's thermal acoustic waves, in good agreement with the formula of Brillouin in (1). The weak signal in these experiments required very careful purification of the samples and exposure times of the photographic plates for upwards of eight hours in some cases.

2.4 Scattering by statistical fluctuations

Mandelstam's paper made one other important observation that Brillouin did not address. Mandelstam recognized that a time-varying disturbance in the medium could take forms other than wave motion. Statistical fluctuations of temperature or entropy continually arise and dissipate in a *diffusive* fashion. He shows that rather than inducing harmonic oscillations, such fluctuations should give rise to an exponentially decaying component to the scattered power $J = J_0 e^{-2\alpha t}$ with

$$\alpha = 4Dk^2 \sin^2\frac{\theta}{2}, \tag{4}$$

where $k = 2\pi/\lambda$ is the free-space wavenumber of the incoming light, and D the relevant diffusion constant. Thus, distinct from the doublet of inelastically scattered lines due to the elastic waves, this contribution is expected to produce a Lorentzian broadening of the elastically scattered Rayleigh line with full-width half maximum (FWHM) $\delta\omega = \alpha$. For entropic fluctuations in a fluid, the appropriate form Boyd (2008) is $D = \kappa/(\rho c_p)$ where κ is the thermal conductivity and c_p the specific heat at constant pressure. For fully backscattered light in the visible, the resulting linewidth is of the order of 10 MHz for a liquid such as CS_2 at room temperature.

[e] Feinberg (2002) recalls that Mandelstam had been collaborating with Gross and might have expected to share in the publication, but this issue is down-played in an interesting article by Novikov (2000).

The presence of this unshifted line was noted by Gross and others in some of the early observations of Brillouin scattering in fluids, but curiously Mandelstam's prediction of this feature seems to have been almost completely overlooked, with the credit invariably going to Landau and Placzek (1934), who worked out the detailed theory in 1934.

3. Evolution of Brillouin scattering

Since the first experimental observations of Brillouin scattering, there have been four overall eras within the field, corresponding to the main platforms in which the effect was measured and harnessed. The first era immediately followed the experiments of Gross, and involved somewhat painstaking scattering experiments in bulk materials and in liquid and gas cells; these observations were limited by the broad linewidth of the sources available and consequent weakness of the effect. This era ended abruptly with the development of the laser, which enabled rapid, accurate Brillouin measurements. Shortly thereafter, the invention of the optical fiber meant that Brillouin scattering became important as a tool for sensing and other applications, as well as putting limits on the power carrying-capacity of fibers. The fiber-Brillouin era held sway until recently, when new fabrication technologies have enabled tuning and enhancement of the Brillouin response through the use of nanostructured waveguides and fibers. We cover these four eras briefly here before moving on to the basic principles of Brillouin scattering in the following section.

3.1 Prelaser studies

The measurement of Brillouin scattering prior to the invention of the laser presented considerable challenges. An excellent and informative review of these difficulties, as well as a historical overview of the prelaser Brillouin experiments is given by Rank (1971), who was instrumental in adapting and improving the method in the 1930s and onwards. Much of the motivation of this work was to measure the Landau–Placzek ratio, which is the ratio of the integrated line intensities from Rayleigh scattering versus Brillouin scattering, and was known to be related to the specific heat of the material (Landau and Placzek, 1934). A comprehensive history of the prelaser Brillouin era can be found in Rank's review (Rank, 1971). The difficulties involved in these experiments prior to the laser are nicely summed up in a comment by Benedek and Greytak in a 1965 review:

Until recently, however, the full promise of this technique was not fulfilled because of the lack of monochromaticity and the relatively low intensity of conventional light sources. The breadth of the spectrum of the incident light made it difficult to resolve the Brillouin doublets and almost impossible to determine their natural line width. The advent of the laser has radically altered this state of affairs.

Benedek and Greytak (1965)

3.2 The laser revolution

The first observation of Brillouin Scattering using a strong coherent source was reported by Raymond Chiao, Charles Townes, and Boris Stoicheff in 1964, shortly after the invention of the laser (Chiao et al., 1964). In this experiment (see Fig. 1), the light from a ruby "maser" (with a wavelength of 694 nm) was directed into samples of quartz and sapphire and the back-scattered light mixed with the laser light via a beam-splitter, the output of which was directed through a Fabry-Perot interferometer onto photographic plates (see Fig. 1) which revealed strong Brillouin-shifted lines. The high optical input intensity enabled by a laser facilitated a completely new regime of strong opto-acoustic feedback in the Brillouin process. This fully nonlinear interaction produced much stronger Stokes lines, as reflected in the new term in the title of their paper, *stimulated* Brillouin scattering.

This pioneering work was followed soon after by observations of Brillouin scattering in liquids by Garmire and Townes (1964), who used a similar setup with a liquid cell containing CS_2. In this study the authors noticed that the scattered Brillouin lines were still too weak to be detected directly, however, by redirecting the back-reflected light through the ruby

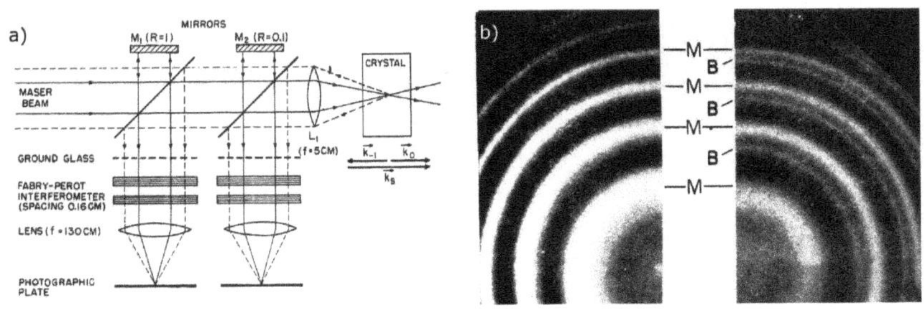

Fig. 1 (A) Experimental setup of the original SBS experiment from Chiao et al. (1964). Light was directed from a ruby laser into the sample. Backscattered light could be distinguished from the laser lines by inserting a mirror with reflectivity $R = 1$ into the setup, allowing direct readout of the laser lines on the plate. The resulting exposures are shown in (B): The left exposure shows the laser lines with the M_1, $R = 1$ mirror inserted. The right-hand side of (B) shows the Brillouin bands appearing faintly between the laser lines.

laser crystal they found that the Stokes signal could be amplified and the Brillouin lines observed. These experiments were followed by observations of laser-induced Brillouin (and Raman) scattering in gases (Hagenlocker and Rado, 1965). A comprehensive overview of the development of laser-induced Brillouin scattering can be found in Elsa Garmire's review (Garmire, 2017).

The invention of the laser gave new impetus to the use of Brillouin scattering for measuring mechanical material properties. Only a month after the original experiment by Chiao et al., the team of Benedek and Greytak (1965) used Brillouin scattering from a range of solid and liquid samples to measure sound speeds. Unlike the backscattered experiments, these experiments involved spectral measurements of side-scattered light from a helium-neon laser. By the early 1970s, Brillouin spectroscopy was well established as the standard for the measurement of elastic and photoelastic properties of materials (see for example the review of Vacher and Boyer (1972)).

3.3 Brillouin scattering in optical fibers

One of the key developments in Brillouin Scattering was the harnessing of the effect within an optical waveguide, rather than in a bulk solid or a fluid/gas cell. The first waveguide experiment was performed in 1972 by Ippen and Stolen (1972). This measurement used 18.5 m of single-mode glass fiber, with a pulsed laser source and a very similar setup to the original Chiao, Townes, and Stoicheff experiment. Observing a low threshold for the measurement, the authors remarked that SBS "limits the amount of narrow-band power which one can transmit through a fiber," thereby anticipating a key motivation for much Brillouin research in the 1980s. Due to the strong optical and acoustic confinement enabled by the fiber geometry, this experiment brought the field into the regime of strong optical gain for the Stokes field, and thus the era of strong SBS.

The ability to easily generate and control narrow-bandwidth light led to new developments in the types of Brillouin effects that can be observed. In a pair of papers in 1985 (Shelby et al., 1985a,b), Shelby, Levinson, and Bayer demonstrated that Brillouin scattering could also occur in the forward direction, mediated by transverse acoustic modes that are trapped by the finite dimensions of the fiber. This effect, known as Guided Acoustic Wave Brillouin Scattering or GAWBS in the context of fibers, paved the way for "forward Brillouin scattering" (FBS) measurements, first in photonic crystal fibers (see chapter "Brillouin scattering in photonic crystal fibers" by Dainese et al.) and then in integrated optical waveguides (see chapter "On-chip stimulated Brillouin scattering" by Pant et al.)

The harnessing of Brillouin scattering in optical fibers led to a number of significant developments in Brillouin applications. The fact that Brillouin effects occur at relatively low power levels in fiber means that they can be used as a source of narrow-band gain. This observation resulted in the first Brillouin laser—a ring fiber laser—built by Hill, Kawasaki, and Johnson in 1976 (Hill et al., 1976). SBS-based fiber ring lasers were in turn used for such new applications as gyrotropic sensing (Thomas et al., 1980). The use of SBS in fibers for distributed sensing of quantities such as temperature (using thermally induced changes in the Brillouin shift) was proposed in Culverhouse et al. (1989). At the same time the use of SBS for optical signal processing was beginning to be explored, notably in the new field of microwave photonics (see the review by Loayssa et al. (2002) or chapter "SBS for Microwave Photonics (MWP)", Liu et al., contained in Part 2 of this volume).

As noted by Ippen and Stolen, the presence of SBS also put significant limits on the power-carrying ability of optical fibers at narrow linewidths. This is because for sufficiently long fiber lengths, or sufficiently high optical powers, the majority of the light in a narrow-linewidth beam would undergo Brillouin scattering and be back-reflected. A widely used model for the power-carrying capacity due to Brillouin scattering was proposed by Smith in 1972 (Smith, 1972). Considerable research in the 1970s and 1980s was addressed to solving this problem (see, for example, the reviews of Cotter (1983) and Garmire (2017)).

For an excellent review of the physics of SBS in optical fibers the reader is directed to the article by Kobyakov et al. (2010). Brillouin scattering is also well covered in the standard textbooks on nonlinear optics, in particular those of Boyd (2008) and Agrawal (2012). The 2017 review of Garmire (2017) provides an excellent overview not only of the early days of laser-induced Brillouin experiments but also goes into the detail of the development of Brillouin fiber lasers and related applications.

3.4 Brillouin scattering in microstructured and integrated waveguides

In the first decade of the 21st century, new fabrication techniques in photonics began to enable control over opto-acoustic interactions to an unprecedented degree. This was demonstrated in several significant advances (e.g., cooling to the ground state) in resonator-based optomechanics (see chapter "The convergence of cavity optomechanics and Brillouin scattering" by Schmidt et al.) It was soon realized that many of these advances—in particular the use of nanofabrication techniques to control the acoustic response—could also be applied to optical waveguides.

In 2006 it was demonstrated that photonic crystal fibers (PCFs) could be used to control the Brillouin spectrum (Dainese et al., 2006); this initial result was soon followed by a demonstration that SBS could be greatly enhanced using a microstructured fiber core, suspended by thin strands and thereby tightly confining both optical and acoustic modes (Kang et al., 2008).

The observation and harnessing of Brillouin scattering in integrated photonic circuits presented considerable challenges. First among these is that the simultaneous confinement of light and sound waves within a waveguide core is not straightforward. PCFs confine the sound using cores that are isolated from the cladding by thin strands, however, such isolation is considerably more difficult to achieve in on-chip waveguides. One alternative is to find a compatible pair of materials for which both light and sound may be confined to the waveguide core using total internal reflection; such materials must have the phase speed in the core lower than in the cladding for both light and sound waves. This was precisely the strategy followed in the first demonstration of on-chip Brillouin scattering (Pant et al., 2011), in which the optical and acoustic modes were confined together in a soft, high-index chalcogenide glass core. New theoretical work appearing around this time (Rakich et al., 2012) demonstrated that extremely high gains could be obtained in microstructured waveguides by exploiting "moving boundary" (Johnson et al., 2002) and radiation pressure effects. This work motivated an intense effort to harness strong on-chip Brillouin effects, with the first demonstrations in suspended silicon waveguides (Shin et al., 2013), followed by elegant experiments in waveguides balanced on pedestals (Van Laer et al., 2015b) and suspended between supports (Van Laer et al., 2015a). In these structures, the necessary acoustic confinement is provided not by total internal reflection, which is impossible in the silicon-on-insulator platform, but by reducing acoustic leakage into the substrate through the narrow support structures to negligible levels. These groundbreaking experiments formed the platforms that are the basis for much of the work described throughout this book.

4. Basic principles of Brillouin scattering

Let us now turn to the present, and provide a high-level introduction to the basic elements of SBS in several configurations. The full theory will be developed in chapter "Theoretical formalisms for stimulated Brillouin scattering" by Wolff et al.

We consider a waveguide oriented along the z-axis (see Fig. 2), which carries a continuous optical "pump" wave of frequency ω_p and wavenumber

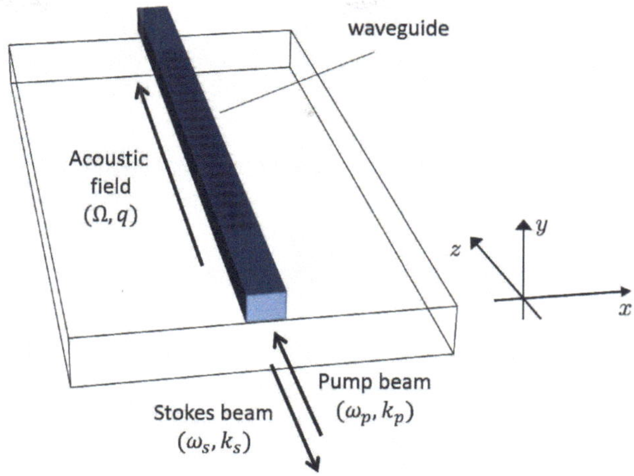

Fig. 2 Illustration of a simple Brillouin interaction: An optical pump is input to one end of an optical waveguide; acoustic waves in the waveguide scatter the optical field into an output Stokes wave.

k_p traveling in the positive z direction. Brillouin scattering occurs when this optical wave is scattered from acoustic waves in the waveguide. These acoustic waves may come from thermal phonons, as discussed by Brillouin in his original paper, however, they can also be externally driven, e.g., by a piezoelectric device (Sohn et al., 2018), and in the most common forms of Brillouin scattering the acoustic waves are excited optically. There are several underlying physical effects that can cause this scattering: first, an acoustic wave compresses the material, changing the permittivity of the waveguide via *photoelasticity*. Second, the acoustic wave will deform any physical boundaries of the medium, leading directly to a change in the permittivity at the waveguide edges and any internal interfaces. Finally other effects, including *roto-opticity*, may arise if the waveguide consists of a material that is optically anisotropic; these effects are discussed in Section 3.4.4 of chapter "Theoretical formalisms for stimulated Brillouin scattering" by Wolff et al. Each of these mechanisms results in a periodic change in the optical properties of the waveguide (i.e., a grating) moving at the speed of sound corresponding to the frequency of the acoustic wave. As with any grating, light will be efficiently scattered if the relevant phase-matching conditions are satisfied.

If the acoustic wave has frequency Ω and wavenumber q, then the light will be scattered into a wave of frequency ω_s and wavenumber k_s, where

$$\omega_s = \omega_p - \Omega, \qquad k_s = k_p - q \,. \tag{5}$$

The frequency Ω of the acoustic wave determines the separation of the Stokes from the pump and is known as the *Brillouin shift*. Equations (5) are phase-matching conditions that ensure the conservation of energy and momentum in the scattering process. Given that the acoustic wave induces a moving grating with velocity $v_{ac} = \Omega/q$, then Eqs. (5) represent the Doppler shift and momentum change that the light undergoes on reflection (or forward scattering) from this grating.

On a fundamental level, the scattering of light from sound waves involves the creation and annihilation of photons and phonons. If a phonon is created in the scattering process, as must happen to conserve energy and momentum when the light is scattered from a grating, then this is known as the Stokes process and leads to a photon with downshifted frequency as in Eqs. (5). However, a phonon can also be absorbed in the interaction—this is known as the anti-Stokes process, and results in the creation of photons with frequency higher than the pump. Both Stokes and anti-Stokes processes occur in Brillouin scattering, however, we shall see in the following sections that the Stokes and anti-Stokes lines can become unbalanced by the presence of gain.

For the right combination of wavevectors and frequencies appearing in Eqs. (5), light will be efficiently scattered from the pump into another guided optical mode. There are in general three possibilities for this scattering, which lead to three classes of Brillouin processes, each with its own distinct features and dynamics. In *Backward Brillouin Scattering*, the optical mode is reflected directly backwards, into the backward-propagating version of the pump's own spatial mode; in Forward Brillouin Scattering (FBS), light from the pump is scattered in the forwards direction into the same mode as the pump but at a shifted frequency. If the waveguide is multimoded then light can be scattered into an entirely different mode to the pump, a process known as *Intermodal Brillouin scattering* (IBS). The type of scattering that occurs can be determined from dispersion diagrams showing both the optical and acoustic modes (see Figs. 3A, 4A, and 5A): fulfillment of the phase-matching conditions in Eqs. (5) requires a closed vector sum on these diagrams between the optical and acoustic modes participating in the interaction. In the following we discuss the important features of each of these main Brillouin processes.

4.1 The backward Brillouin process and stimulated Brillouin scattering

Scattering in the backwards direction arises naturally from the interaction between light and thermal phonons, as predicted by Brillouin and

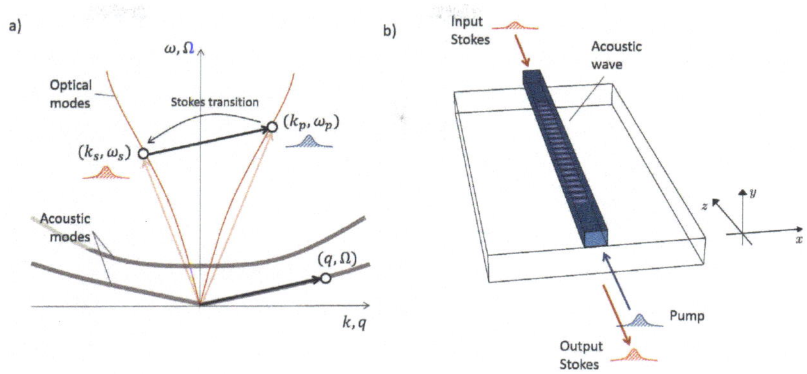

Fig. 3 (A) Illustration of phase-matching in backward Brillouin scattering. Note that the slopes and frequencies of the acoustic curves have been greatly exaggerated. (B) Schematic of optical interactions in backwards Brillouin scattering.

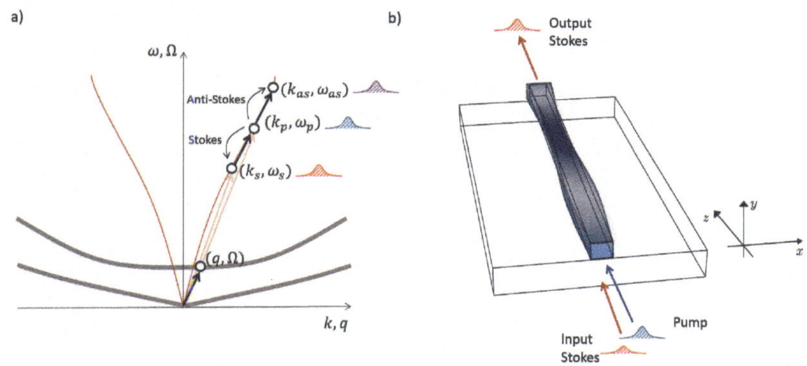

Fig. 4 (A) Illustration of phase-matching in forward Brillouin scattering; (B) Schematic of optical interactions in forwards Brillouin scattering.

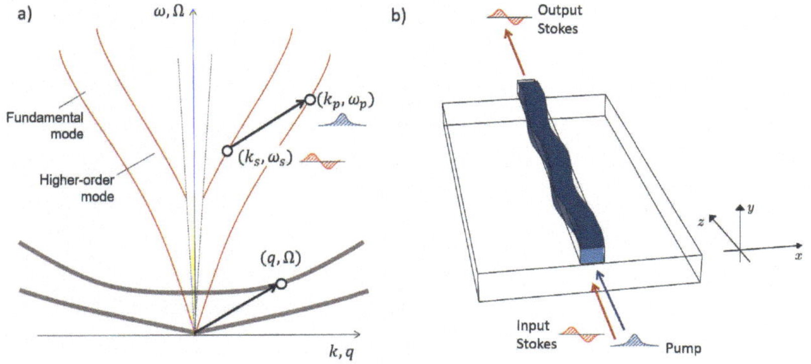

Fig. 5 (A) Phase-matching diagram for intermodal Brillouin processes; (B) schematic of wave interactions in intermodal Brillouin scattering.

Mandelstam. For back-scattering to occur, the momentum of the back-scattered light, given by the wavenumber $k_s \approx -k_p$ (see Fig. 3), must be compensated by a generated acoustic wave with $q \approx 2k_p$ according to the second equation in (5). Thus the acoustic waves participating in the interaction should have approximately half the wavelength of the optical waves in the waveguide. These acoustic modes typically have dominant strain components that are aligned along the propagation direction (Fig. 3B), and so resemble pressure waves with phase fronts transverse to the z-axis. Such modes are often called *longitudinal* modes, a terminology that refers both to the alignment of the motion along the waveguide axis as well as to the fact that they are predominantly pressure waves, which are purely longitudinal in bulk materials, as opposed to shear waves, which are transverse. Because the velocity of shear waves is always lower than that of pressure waves, it is possible to find combinations of materials where the phase velocity of a mode localised to the waveguide core lies below the speed of the shear waves in the cladding; in these cases acoustic guidance via total internal reflection is possible.

The dispersion curve of such an acoustic mode is a straight line with slope given approximately by the longitudinal acoustic velocity v_{ac}, and so to good approximation, the Brillouin shift is

$$\Omega = qv_{ac} \approx 2k_p v = 2n_{eff}\frac{\omega_p v_{ac}}{c}. \tag{6}$$

Here the acoustic waves are typically predominantly pressure waves, and so the velocity is given by $v_{ac} = \sqrt{M/\rho_0}$, where M is the longitudinal wave modulus of the material. Note that the relation (6) accords with the original Brillouin relation (3) for scattering through the angle $\theta = \pi$. This family of acoustic modes produces negligible displacements on the waveguide boundaries, and so the dominant mechanism for scattering from these waves is photoelasticity.

An interesting and important effect, known as *Stimulated Brillouin Scattering (SBS)*, occurs in the backward scattering process. This effect can be understood in the following way: any backscattered light will interfere with the forward-propagating pump, creating a moving optical intensity pattern, which in turn creates a grating in the material via photoelasticity and the moving boundary effect. The speed of this grating is equal to the ratio of the difference in frequency of the optical waves to the difference in their wavenumbers, so for a pump with parameters (ω_p, k_p) and a scattered wave with (ω, k), the velocity of the grating is $v = (\omega_p - \omega)/(k_p - k)$. If this velocity v is exactly equal to the acoustic phase velocity in the material v_{ac}, then an acoustic wave is resonantly excited in the waveguide; the optical frequency at which this occurs is the Stokes frequency, given by the phase-matching condition (5). Because it it resonantly excited, the grating associated with the acoustic wave then becomes very strong, and because the grating is moving, it will reflect further light from the pump, Doppler-shifted to the Stokes frequency, which then further amplifies the acoustic wave, and so on. This strong self-amplifying behavior of both the Stokes and acoustic wave results in exponential growth of the Stokes wave at the expense of the pump, in other words, *optical gain*. The result of this is an extremely strong backscattered Stokes wave, downshifted from the pump by a shift of typically around 5–20 GHz, and with a linewidth of 10–100 MHz.

On a fundamental level, the gain in SBS is caused by the creation of phonons, which is stimulated by the preexisting phonon field. This process can be "kicked off" by the inherent thermal fluctuations in the waveguide, upon which the Stokes is amplified. For sufficiently long waveguides, and for pump linewidths smaller than the Brillouin linewidth, all of the pump is converted into a counter-propagating Stokes wave. It should be noted that the anti-Stokes process involves the annihilation of phonons and therefore does not experience gain. Measurement of the Brillouin spectrum with a weak pump will therefore exhibit both Stokes and anti-Stokes Brillouin lines, however, the Stokes line will quickly dominate as the pump power is increased.

For time-independent pump and Stokes waves, the evolution of the power of the pump and Stokes is determined by the equations (see chapter "Theoretical formalisms for stimulated Brillouin scattering" by Wolff et al., Section 4.1):

$$\begin{aligned} \frac{dP_p}{dz} &= -g_{\mathrm{SBS}} P_p P_s - \alpha P_p \\ \frac{dP_s}{dz} &= -g_{SBS} P_s P_p + \alpha P_s \end{aligned} \tag{7}$$

where α is the optical power loss in m^{-1}, and g_{SBS} is the Brillouin power gain measured in $\mathrm{m}^{-1}\mathrm{W}^{-1}$, which depends on the overlap of the acoustic field with the optical field, as well as on the optical and acoustic material properties and any possible detuning of the input Stokes field from the peak of the

Brillouin response. If the pump intensity is much larger than the Stokes then it will remain undepleted except by loss. In this case, it can be shown that the amplification of the Stokes from its initial value at $z = L$ to the output at $z = 0$ is

$$P_s(0) = P_s(L) \exp\left[g_0 P_p(0) L_{\text{eff}} - \alpha L\right] \tag{8}$$

where $L_{\text{eff}} = (1 - e^{-\alpha L})/\alpha$ is the usual expression for the effective length of the waveguide (Agrawal, 2012).

SBS can be used to amplify a counter-propagating input Stokes beam, however, the effect can also be triggered by thermal phonons in the absence of a Stokes seed. For a sufficiently strong pump with a linewidth below the Brillouin linewidth, and provided the waveguide length is long enough, SBS can result in the back-reflection of the entirety of the pump. This can create significant problems in high-power lasers and other devices that rely on the transmission of large amounts of optical power over narrow linewidths. On the other hand, because the gain is proportional to the existing intensity of the pump field, the backward process behaves like a very strong third-order nonlinearity, with a threshold around two orders of magnitude smaller than the thresholds for Raman amplification or the parametric gain induced by the Kerr effect (Agrawal, 2012).

4.2 The forward Brillouin process

In FBS (Fig. 4), an acoustic mode with small wavenumber $q \approx 0$ at a nonzero frequency Ω scatters light from the pump into a nearby sideband of the optical mode. The small wavenumber of the acoustic mode means that the acoustic field propagates almost directly transverse to the waveguide axis, and so FBS is facilitated by acoustic modes that are close to being transverse acoustic resonances of the waveguide. Because these modes are almost "flat" at the $q = 0$ point (see Fig. 4A), the Brillouin shift for FBS occurs at specific frequencies that depend only on the mechanical material properties and geometry of the waveguide, rather than on the frequency of the pump and Stokes waves, as in the backwards process. A good approximation for the Brillouin frequency for FBS coupling to the first-order breathing mode of a rectangular waveguide of width a is (Wolff et al., 2021)

$$\Omega \approx \frac{\pi}{a} \sqrt{\frac{E}{\rho}} \tag{9}$$

where E is the Young's modulus and ρ is the material density of the core.[f] Because the dominant motion of the modes in FBS modes is perpendicular to the waveguide axis, such modes are highly leaky if they are not suspended or otherwise isolated from their environment. The interaction tends to be highly determined by radiation pressure and by the moving boundary effect, although electrostriction and photoelasticity can also play an important role (Rakich et al., 2012).

Unlike SBS, FBS does not result in gain. This is because, provided the optical dispersion is sufficiently weak, the process is symmetric with respect to the generation of Stokes and anti-Stokes photons: every phonon that is generated in producing a Stokes photon with frequency $\omega_s = \omega_p - \Omega$, is, on average, absorbed in the generation of anti-Stokes photons with frequency $\omega_{as} = \omega_p + \Omega$. The acoustic field therefore cannot grow exponentially as happens in SBS; instead it achieves a static value proportional to the total optical energy at each point in z. This acoustic field then couples the pump with the Stokes and anti-Stokes fields, as well as with higher-order Stokes and anti-Stokes fields. This causes energy to be shuffled between the various orders in the same way as occurs in an opto-acoustic modulator: the Stokes grows along the first part of the waveguide before eventually losing energy to the pump and to successively higher-order Stokes fields (Wolff et al., 2017).

4.3 The intermodal process

In IBS, light is scattered from the pump into a different spatial mode of the waveguide (Fig. 5). This process can occur either in the forward or the backward direction. Because there is no symmetry between the Stokes and anti-Stokes processes (as there is in FBS) such processes can experience gain in a similar way to backward SBS. The main difference to SBS is that the acoustic modes need not be predominantly longitudinal—in particular the forward intermodal scattering processes may feature acoustic modes with strongly hybrid transverse and longitudinal nature. In addition the process requires strict conditions on the relative symmetries of the acoustic and optical modes: for example a flexural acoustic mode is needed to couple between the fundamental and first-order optical modes (Wolff et al., 2014).

[f] Here the Young's modulus E is used because it describes the response of strain to compression in a linear direction—in this case along the width of the waveguide. This is different to the bulk modulus, which describes the relative change in volume (dilatation) in response to isotropic compression, and is also different to the longitudinal modulus M, which describes the speed of pressure waves that move through the material.

4.4 On the precise meaning of spontaneous and stimulated Brillouin scattering

Within the literature there are some differences between what is typically meant by the labels "spontaneous" and "stimulated" in the context of Brillouin Scattering, and what these terms usually mean in quantum mechanics and in the theory of lasers. This is worth exploring so as to avoid confusion in the chapters ahead.

There is a tradition in the Brillouin community of using the words "spontaneous" and "stimulated" to distinguish between different types of Brillouin experiment, rather than directly identifying the process itself. A "spontaneous" Brillouin measurement often means one that arises from an experiment where there is no externally input Stokes field, and in which the Brillouin process is therefore initiated by the thermal background. A "stimulated" measurement is one that is obtained by measuring the gain experienced by an input Stokes field. Typically one would measure the "spontaneous" field first in order to get an idea of where the Brillouin frequency shift is, followed by a "stimulated" measurement that applies a narrow-linewidth Stokes source near this value to measure the spectrum.

The use of "stimulated" in this sense refers to the amplification of a signal, and so there is a well-defined connection with the idea of stimulated processes in lasers: the self-reinforcing behavior that we see in backwards Brillouin scattering is a macroscopic manifestation of the fact that the probability of a fundamental interaction—the annihilation of a pump photon, together with the creation of a Stokes photon and a phonon—is increased in the presence of existing Stokes photons. This leads to the observed gain, and so any Brillouin process that results in gain can be referred to as SBS with some justification. One must, however, be careful when discussing FBS: one may find reference to a "stimulated" measurement of a process that does not experience gain, and for which there is ultimately no stimulated process involved.

Further confusion may then arise when discussing "spontaneous" measurements. First, a spontaneous process in laser physics and quantum optics normally refers specifically to a transition between states occurring as a result of the coupling to the vacuum field. This is not the case here, where it refers to the coupling of the pump to the bath of (randomly phased) thermal phonons in the waveguide. This difference could be manageable if it were not for the fact that, once initiated, the spontaneous signal can then itself experience gain. If the two terminologies are inadvertently mixed, one finds oneself talking of "stimulated spontaneous Brillouin scattering" to describe the gain experienced by the Stokes in a "spontaneous" measurement. This can be confusing for speaker and listener alike.

One way to avoid confusion of this type is to clearly distinguish the initiation of the process from the gain. One can then speak of "thermally seeded" Brillouin scattering as opposed to a "coherently seeded" measurement; any process that experiences gain is then stimulated Brillouin scattering or SBS. Alternatively, one can retain the "spontaneous" designation for the thermally seeded process, and when such a process experiences gain describe it as *amplified spontaneous Brillouin scattering* (Thévenaz, 2011). This terminology recalls the equivalent effect observed in erbium-doped fiber amplifiers. In any case readers should be aware of the different uses of these terms throughout the literature and within the diverse chapters of this book.

Acknowledgments

C.G.P. and M.J.S. acknowledge the support of the Australian Research Council under Discovery Projects DP160101691, DP200101893 and DP220100488. G.S.W. acknowledges funding by São Paulo Research Foundation (FAPESP) through grants 2018/15577-5 and 2018/25339-4, CNPq and Coordenação de Aperfeiçoamento de Pessoal de Nível Superior—Brasil (CAPES) (Financial Code 001).

References

Agrawal, G.P., 2012. Nonlinear Fiber Optics, fifth ed. Academic Press, p. 529.

Benedek, G., Greytak, T., 1965. Brillouin scattering in liquids. Proce. IEEE 53 (10), 1623–1629.

Boyd, R.W., 2008. Nonlinear Optics, third ed. Academic Press.

Brillouin, L., 1914. Diffusion de la lumière par un corps transparent homogène. Comptes Rendu Séances Acad. Sci. 158, 1331–1334.

Brillouin, L., 1922. Diffusion de la lumière et des rayons X par un corps transparent homogène. Ann. Phys. 9 (17), 88–122.

Brillouin, L., 1969. Birth and growth of the Brillouin scattering. In: Wright, G.B. (Ed.), Light Scattering Spectra of Solids. Springer-Verlag, New York, pp. xvii–xviii.

Burton, H.E., 1945. The optics of Euclid. J. Opt. Soc. Am. 35, 357–372.

Chiao, R.Y., Townes, C.H., Stoicheff, B.P., 1964. Stimulated Brillouin scattering and coherent generation of intense hypersonic waves. Phys. Rev. Lett. 12, 592–595.

Cotter, D., 1983. Stimulated Brillouin scattering in monomode optical fiber. J. Opt. Commun. 4, 10–19.

Culverhouse, D., Farahi, F., Pannell, C.N., Jackson, D.A., 1989. Potential of stimulated Brillouin scattering as sensing mechanism for distributed temperature sensors. Electron. Lett. 25, 913–915.

Dainese, P., Russell, P.S.J., Joly, N., Knight, J.C., Wiederhecker, G.S., Fragnito, H.L., Laude, V., Khelif, A., 2006. Stimulated Brillouin scattering from multi-GHz-guided acoustic phonons in nanostructured photonic crystal fibres. Nat. Phys. 2 (6), 388–392.

Debye, P., 1912. Zur Theorie der spezifischen Waerme. Ann. Phys. 39, 789–839.

Einstein, A., 1907. Planck's theory of radiation and the theory of specific heat. Ann. Phys. 22, 180–190.
Einstein, A., 1910. The theory of the opalescence of homogeneous fluids and liquid mixtures near the critical state. Ann. Phys. 33, 1275–1298.
Feinberg, E.L., 2002. The forefather (about Leonid Isaakovich Mandelstam). Uspekhi Fizicheskih Nauk 172 (1), 91.
Garmire, E., 2017. Perspectives on stimulated Brillouin scattering. New J. Phys. 19, 011003.
Garmire, E., Townes, C.H., 1964. Stimulated Brillouin scattering in liquids. Appl. Phys. Lett. 5, 84.
Gross, E., 1930a. Change of wavelength of light due to elastic heat waves at scattering in liquids. Nature 126 (3171), 201–202.
Gross, E., 1930b. Splitting of the frequency of light scattered by liquids and optical anisotropy of molecules. Nature 126 (3181), 603–604.
Gross, E., 1930c. The splitting of spectral lines at scattering of light by liquids. Nature 126 (3176), 400.
Hagenlocker, E.E., Rado, W.G., 1965. Stimulated Brillouin and Raman scattering in gases. Appl. Phys. Lett. 7, 236–238.
Hill, K.O., Kawasaki, B.S., Johnson, D.C., 1976. cw Brillouin laser. Appl. Phys. Lett. 28, 608.
Ippen, E.P., Stolen, R.H., 1972. Stimulated Brillouin scattering in optical fibers. Appl. Phys. Lett. 21, 539.
Johnson, S.G., Ibanescu, M., Skorobogatiy, M.A., Weisberg, O., Joannopoulos, J.D., Fink, Y., 2002. Perturbation theory for Maxwell's equations with shifting material boundaries. Phys. Rev. E 65 (6), 066611.
Kang, M.S., Brenn, A., Wiederhecker, G.S., Russell, P.S.J., 2008. Optical excitation and characterization of gigahertz acoustic resonances in optical fiber tapers. Appl. Phys. Lett. 93 (13), 131110.
Kobyakov, A., Sauer, M., Chowdhury, D., 2010. Stimulated Brillouin scattering in optical fibers. Adv. Opt. Photonics 2 (1), 1–59.
Landau, L.D., Placzek, G., 1934. Structure of the unshifted scattering line. Z. Sowjetunion 5, 172.
Landsberg, G., Mandelstam, L., 1928. Über die Lichtzerstreuung in Kristallen. Z. Phys. 50 (11–12), 769–780.
Lebedev, 1901. Experimental examination of light pressure. Ann. Phys. 6, 433.
Loayssa, A., Benito, D., Garde, M.J., 2002. Applications of optical carrier Brillouin processing to microwave photonics. Opt. Fiber Technol. 8, 24–42.
Mandelstam, L.I., 1908. On optically homogeneous and turbid mediums. In: Complete Works, vols. 1–5. Izd. AN SSSR, pp. 1947–1955.
Mandelstam, L.I., 1926. Light scattering by inhomogeneous media. Zh. Russ. Fiz-Khim. 58, 381–386.
Needham, J., 1974. Science and Civilisation in China. vol. 4 Cambridge University Press.
Nichols, E.F., Hull, G.F., 1903. The pressure due to radiation. Proc. Am. Ac. Arts Sci. 38, 559–599.
Novikov, B.V., 2000. On the experimental discovery of Mandelstam-Brillouin scattering. Physics-Uspekhi 43 (1), 105–107.
Pant, R., Poulton, C.G., Choi, D.Y., Mcfarlane, H., Hile, S., Li, E., Thevenaz, L., Luther-Davies, B., Madden, S.J., Eggleton, B.J., 2011. On-chip stimulated Brillouin scattering. Opt. Express 19 (9), 388–392.
Pechenkin, A., 2019. L.I. Mandelstam and His School in Physics, second ed. Springer, p. 261.
Rakich, P.T., Reinke, C., Camacho, R., Davids, P., Wang, Z., 2012. Giant enhancement of stimulated Brillouin scattering in the subwavelength limit. Phys. Rev. X 2 (1), 011008.
Raman, C.V., 1928. A new radiation. Indian J. Phys. 2. 387–298.

Raman, C.V., Krishnan, K.S., 1928a. A new type of secondary radiation. Nature 121 (3048), 501–502.
Raman, C.V., Krishnan, K.S., 1928b. The optical analogue of the Compton effect. Nature 121 (3053), 711.
Rank, D.H., 1971. Brillouin effect in liquids in the prelaser era. J. Acoust. Soc. Am. 49, 937–940.
Rashed, R., 1990. A pioneer in anaclastics: Ibn Sahl on burning mirrors and lenses. Isis 81, 464–491.
Rayleigh, 1899. XXXIV. On the transmission of light through an atmosphere containing small particles in suspension, and on the origin of the blue of the sky. Lond. Edinb. Dublin Philos. Mag. J. Sci. 47 (287), 375–384.
Shelby, R.M., Levenson, M.D., Bayer, P.W., 1985a. Guided acoustic-wave Brillouin scattering. Phys. Rev. B 31, 5244.
Shelby, R.M., Levenson, M.D., Bayer, P.W., 1985b. Resolved forward Brillouin scattering in optical fibers. Phys. Rev. Lett. 54, 939–942.
Shin, H., Qiu, W., Jarecki, R., Cox, J.A., Olsson, R.H., Starbuck, A., Wang, Z., Rakich, P.T., 2013. Tailorable stimulated Brillouin scattering in nanoscale silicon waveguides. Nat. Commun. 4, 1944.
Singh, R., Riess, F., 2001. The 1930 Nobel Prize for Physics: a close decision? Notes Rec. R. Soc. Lond. 55 (2), 267–283.
Smith, R.G., 1972. Optical power handling capacity of low loss optical fibers as determined by stimulated Raman and Brillouin scattering. Appl. Opt. 11 (11), 2489–2494.
Smith, A.M., 1996. Ptolemy's Theory of Visual Perception: An English Translation of the Optics. American Philosophical Society, p. 300.
Smoluchowski, M., 1908. Molekular-kinetische Theorie der Opaleszenz von Gasen in kritisehen Zustande. sowie einiger verwandter Erscheinungen. Ann. Phys. 25, 205–226.
Sohn, D.B., Kim, S., Bahl, G., 2018. Time-reversal symmetry breaking with acoustic pumping of nanophotonic circuits. Nat. Photonics 12 (2), 91–97.
Thévenaz, L., 2011. Inelastic scatterings and applications to distributed sensing. In: Thévenaz, L. (Ed.), Advanced Fiber Optics—Concepts and Technology. EPFL Press, distributed by CRC Press, Lausanne, Switzerland, pp. 263–307.
Thomas, L.H., 1985. Léon Nicolas Brillouin 1889–1969. Natl. Acad. Sci. 55, 69–92.
Thomas, P.J., van Driel, H.M., Stegeman, G.I.A., 1980. Possibility of using an optical fiber Brillouin ring laser for inertial sensing. Appl. Opt. 19, 1906–1908.
Vacher, R., Boyer, L., 1972. Brillouin scattering: a tool for the measurement of elastic and photoelastic constants. Phys. Rev. B 6, 639–673.
Van Laer, R., Bazin, A., Kuyken, B., Baets, R., Van Thourhout, D., 2015a. Net on-chip Brillouin gain based on suspended silicon nanowires. New J. Phys. 17 (11), 115005.
Van Laer, R., Kuyken, B., Van Thourhout, D., Baets, R., 2015b. Interaction between light and highly confined hypersound in a silicon photonic nanowire. Nat. Photonics 9, 199–203.
Wolff, C., Steel, M.J., Poulton, C.G., 2014. Formal selection rules for Brillouin scattering in integrated waveguides and structured fibers. Opt. Expr. 22, 32489.
Wolff, C., Stiller, B., Eggleton, B.J., Steel, M.J., Poulton, C.G., 2017. Cascaded forward Brillouin scattering to all stokes orders. New J. Phys. 19 (2), 023021.
Wolff, C., Smith, M.J.A., Stiller, B., Poulton, C.G., 2021. Brillouin scattering—theory and experiment: tutorial. J. Opt. Soc. Am. B 38 (4), 1243–1269.

CHAPTER TWO

Theoretical formalisms for stimulated Brillouin scattering

Christian Wolff[a,*], Christopher G. Poulton[b], Michael J. Steel[c], and Gustavo Wiederhecker[d]
[a]Center for Nano Optics, University of Southern Denmark, Odense M, Denmark
[b]School of Mathematical and Physical Sciences, University of Technology Sydney, Sydney, NSW, Australia
[c]School of Mathematical and Physical Sciences, Macquarie University, Sydney, NSW, Australia
[d]Gleb Wataghin Physics Institute, University of Campinas, Campinas, SP, Brazil
*Corresponding author: e-mail address: cwo@mci.sdu.dk

Contents

Semiconductors and Semimetals, Volume 109
ISSN 0080-8784
https://doi.org/10.1016/bs.semsem.2022.04.002

1. Introduction

Any theory of Brillouin scattering must describe the behavior of the optical (electromagnetic) waves within a system, the induced acoustic (elastic) waves, as well as the interaction between these two fields. In almost all Brillouin interactions, the coupling between optical and acoustic fields is such that large interaction regions, resonant structures, long periods of time, or some combination of these are needed in order to obtain a measurable Brillouin signal. For example, a typical backward Brillouin experiment involves a cm-scale waveguide with a micron-scale cross section, with interaction times over tens of nanoseconds, and with the acoustic displacements of the waveguide boundaries being of the order of 1 nm. For these and similar situations, it is extremely challenging to numerically solve the full coupled problem for the electromagnetic and elastic waves and so far this has only been demonstrated within some other approximations, such as restriction to two dimensions and to the frequency domain (Shi et al., 2017). On the other hand, the weak coupling results in field envelopes that change very slowly with respect to the optical and acoustic frequencies involved in the interaction. For this reason the natural tool for studying Brillouin interactions is coupled mode theory.

There are several ways to derive the coupled mode equations (CMEs) for Brillouin interactions. The classical approach appears in previous textbooks on nonlinear optics (Agrawal, 2012; Boyd, 2008), and it is possible to extend this to the full vector/tensor formalism needed in modern Brillouin waveguides and resonators (Sipe and Steel, 2016; Wolff et al., 2015b). It is also possible to use a Lagrangian formalism to simulate the interactions (Laude and Beugnot, 2015).

We use here an approach based on the Hamiltonian of the interaction, since this approach gives a direct connection to non-Brillouin optomechanics literature as well as to the quantum-mechanical formulation (Sipe and Steel, 2016). Such an approach also has the advantage of making explicit several approximations that must be made to obtain the CMEs for the slowly varying envelopes of the interacting fields.

We begin here with a brief presentation of the theory of linear elasticity, which may not be familiar to researchers in photonics, and describe the equations for elastic modes in both resonant cavities and waveguides. After a summary of the necessary electromagnetic theory for describing optical modes, we give the appropriate form of the Hamiltonian that describes the behavior of the coupled system. From the Hamiltonian, we derive the CMEs, incorporating the underlying physical effects that give rise to the coupling: photoelasticity, radiation pressure, the roto-optic effect, and so on.

Having expressed the theory in its most general form, we then study three specific configurations that are the basis of many Brillouin experiments: these are backward SBS (BSBS), forward Brillouin scattering (FBS), and intermodal Brillouin scattering. The appropriate form of the CMEs depends on the problem that is being studied: for example, backward Brillouin scattering requires only two optical modes and one acoustic mode in its simplest manifestation, and therefore a set of three equations is needed, whereas FBS can lead to cascaded optical fields and so requires a more generous modal expansion. We present the CMEs for each of these situations, and discuss the characteristic behavior of the solutions. We also discuss alternative formulations for the theory, and the circumstances in which they are either equivalent or perform well. In particular, one can often assume that all fields can be described by scalar quantities, especially in fluids and optical fibers. We link the full theory to this simplified case, and compare other modeling approaches.

In Section 5 we present the formulation for Brillouin scattering in resonant cavities. Again we use the Hamiltonian approach, and link this to the more familiar formalism of optomechanics. We derive the general time evolution equations for resonant Brillouin systems, which will be used at various points throughout this volume. We conclude with a list of limitations of the theory, together with directions to references where the theory has been extended.

2. Linear elasticity and electromagnetism

2.1 Essentials of linear elasticity theory

To begin the development of the theory, let us summarize the foundations of linear elasticity theory. We refer the reader to the excellent texts by Auld (1990a,b) for a comprehensive treatment that is notable for its development of analogies with classical electromagnetism.

Both vector and tensor quantities and notation are ubiquitous in elasticity theory. Throughout the book, vector quantities such as the displacement $\boldsymbol{U}(\boldsymbol{r})$ are denoted with bold font, and vector components with a single subscript, e.g., $U_i(\boldsymbol{r})$. General tensor quantities are denoted with a double-arrow, e.g., $\overleftrightarrow{T}$, and individual components using double subscripts (T_{ij}). While we generally favor equations in index form, we will sometimes use the dot ($\boldsymbol{A} \cdot \boldsymbol{B} \equiv A_i B_i$) and double dot ($\overleftrightarrow{A}:\overleftrightarrow{B} \equiv A_{ij} B_{ij}$) notations to denote single and double tensor contractions. The Einstein convention of an implied sum over repeated indices applies unless otherwise stated.[a]

In the standard theory of continuum classical mechanics, the (real-valued physical) material displacement $\boldsymbol{U}(\boldsymbol{r}, t)$ is governed by Newton's second law (Auld, 1990a)

$$\nabla \cdot \overleftrightarrow{T}(\boldsymbol{r}, t) = \rho_0(\boldsymbol{r}) \frac{\partial^2 \boldsymbol{U}}{\partial t^2} - \boldsymbol{f}(\boldsymbol{r}, t), \tag{1}$$

or in index form

$$\frac{\partial}{\partial r_j} T_{ij}(\boldsymbol{r}, t) = \rho_0(\boldsymbol{r}) \frac{\partial^2 U_i}{\partial t^2} - f_i(\boldsymbol{r}, t), \tag{2}$$

where $\rho_0(\boldsymbol{r})$ is the equilibrium density and $\boldsymbol{f}(\boldsymbol{r}, t)$ represents the volume density of any externally applied body force, and we have used the Einstein summation convention for repeated indices. The *stress* $\overleftrightarrow{T}$ and *strain* $\overleftrightarrow{S}$, both second-rank tensors, are related by the constitutive relation

$$T_{ij} = c_{ijkl} S_{kl} + \eta_{ijkl} \frac{\partial S_{kl}}{\partial t}, \tag{3}$$

where c_{ijkl} (or $\overleftrightarrow{c}$ in index-free notation) is called the *stiffness tensor* and η_{ijkl} is the *viscosity tensor* which leads to damping. These two fourth-rank tensors contain all the material properties of the media under consideration. Finally, the *strain tensor* $\overleftrightarrow{S}$ is defined as the symmetric gradient of the displacement ($\overleftrightarrow{S} = \frac{1}{2}(\nabla + \nabla^t)\boldsymbol{U}$) or in index form

$$S_{ij} = \frac{1}{2}\left(\frac{\partial U_i}{\partial r_j} + \frac{\partial U_j}{\partial r_i}\right). \tag{4}$$

[a] For simplicity in notation, and following the majority of mechanics literature, we make no distinction between covariant (downstairs) and contravariant (upstairs) indices.

Both the stress and strain are symmetric tensors, while the stiffness and viscosity exhibit varying degrees of symmetry depending on the material being considered. The simplest examples are presented in the Appendix, while a comprehensive discussion of the symmetry properties of these tensors for all crystal classes can be found in appendix 2 of Auld (1990a). It is important to be aware that even in isotropic materials (for which the electromagnetic dielectric tensor is a scalar), the stiffness c_{ijkl} depends on two independent quantities, corresponding to the different velocities of longitudinal and shear waves. Indeed, the linear elastic response of a given material is invariably more complicated than the linear dielectric response, which is represented by only a second-rank tensor.

Expressed in terms of the displacement field, Eq. (1) takes the form of the elastic wave equation

$$\frac{\partial}{\partial r_j}\left(c_{ijkl}(\boldsymbol{r})S_{kl}+\eta_{ijkl}\frac{\partial S_{kl}}{\partial t}\right)=\rho_0(\boldsymbol{r})\frac{\partial^2 U_i}{\partial t^2}-f_i(\boldsymbol{r},t), \tag{5}$$

from which, in conjunction with appropriate boundary conditions, one can find the evolution of the elastic field U_i in response to the driving force $f_i(\boldsymbol{r}, t)$.

2.1.1 Modal solutions

In much of this chapter and indeed throughout the two volumes of this collection on Brillouin scattering, we will be concerned with the time evolution of modal expansions of the elastic and electromagnetic fields. Two classes of modes will appear frequently for both types of fields—cavity modes and waveguide modes.

2.1.1.1 Cavity modes

In the context of elastic *cavity* or *microresonator* modes, which arise, e.g., in Section 5 and particularly in all of chapter "The Convergence of cavity optomechanics and Brillouin scattering" by Schmidt et al., one seeks solutions to Eq. (5) of the form

$$\boldsymbol{U}(\boldsymbol{r},t)=\boldsymbol{u}^{(n)}(\boldsymbol{r})e^{-\mathrm{i}\tilde{\Omega}_n t}+\text{c.c.}, \tag{6}$$

where $\boldsymbol{u}^{(n)}(\boldsymbol{r})$ is a complex cavity mode profile, and the tilde on the angular frequency $\tilde{\Omega}_n$ indicates that the frequency may be complex. For these problems, the pairs $\{\boldsymbol{u}^{(n)}(\boldsymbol{r}),\tilde{\Omega}_n\}$ are found as a set of *discrete* eigensolutions to the homogeneous equation

$$\frac{\partial}{\partial r_j}\left(c_{ijkl}(\boldsymbol{r})s_{kl}^{(n)}-i\tilde{\Omega}_n\eta_{ijkl}(\boldsymbol{r})s_{kl}^{(n)}\right)+\tilde{\Omega}_n^2\rho_0(\boldsymbol{r})u_i^{(n)}(\boldsymbol{r})=0, \tag{7}$$

where $s_{kl}^{(n)}(\boldsymbol{r})$ is the modal strain corresponding to $u_n(\boldsymbol{r})$. Solutions are generally found using numerical methods such as finite elements. The modal eigenfrequencies $\tilde{\Omega}_n$ may be complex due to both the intrinsic damping from the viscosity $\overleftrightarrow{\eta}$ and leakage radiation into surrounding cladding or substrate media.

2.1.1.2 Waveguide modes

In contrast, for elastic *waveguide* problems in which the stiffness tensor $\overleftrightarrow{c}(x, y)$ is invariant along the propagation direction z, one seeks propagating modes of the form[b]

$$U(\boldsymbol{r}, t) = \tilde{\boldsymbol{u}}^{(n)}(\boldsymbol{r})e^{-\mathrm{i}\Omega_n t} + \text{c.c.} \tag{8}$$

$$= \boldsymbol{u}^{(n)}(x, y)e^{\mathrm{i}[q_n z - \Omega_n t]} + \text{c.c.}, \tag{9}$$

where n is the band index and we have factored the spatial mode dependence as $\tilde{\boldsymbol{u}}^{(n)}(\boldsymbol{r}) = \boldsymbol{u}^{(n)}(x, y)e^{\mathrm{i}q_n z}$. These states form a series of bands of continuous solutions with real frequency Ω and wavenumber q. At the GHz frequencies relevant for Brillouin physics, it is generally not appropriate to neglect the viscous damping and in principle the wavenumber q should be complex. In most treatments, including ours in this chapter, however, one solves the modal expansion problem for a lossless medium ($\overleftrightarrow{\eta} = 0$) and incorporates the losses through damping terms in the CMEs. One can then treat either q or Ω as the independent variable and eigenvalue respectively. For numerical solutions, there are some advantages to taking the wavenumber q as the free parameter and seeking modes with frequency eigenvalue $\Omega(q)$ (Sturmberg et al., 2018). The eigenvalue equation that follows from Eq. (5) is then

$$-\frac{\partial}{\partial r_j}\left(c_{ijkl}(\boldsymbol{r})\frac{\partial \tilde{u}_k^{(n)}}{\partial r^l}\right) = \rho_0(\vec{r})\Omega^2\, \tilde{u}_i^{(n)}, \tag{10}$$

where we have used symmetry properties of $\overleftrightarrow{c}$ (see Appendix) to eliminate the strain and write an equation purely in terms of the displacement $\tilde{\boldsymbol{u}}^{(n)}(\boldsymbol{r})$. The wavenumber q is implicitly contained in this equations in the z-dependence assumed in Eq. (8); the eigenvalue problem in terms of the transverse mode pattern $\boldsymbol{u}^{(n)}(x, y)$ differs in that all z-derivatives are replaced according to $\partial_z \rightarrow \mathrm{i}q$.

[b] We address the issues of normalization in both Eqs. (6) and (8) in detail below.

Particularly in the context of developing a quantum theory, it is useful to have an expansion in terms of modes of a Hermitian equation (Sipe and Steel, 2016). Notably, this guarantees the possibility of an expansion in terms of a complete orthonormal set of modes. However, the presence of the spatially varying density $\rho_0(\boldsymbol{r})$ in Eq. (10) makes it less obvious how to achieve this. One approach is to introduce an *auxiliary* set of modes defined as

$$\boldsymbol{f}^{(n)}(x,y) = \sqrt{\rho_0(x,y)}\boldsymbol{u}^{(n)}(x,y). \tag{11}$$

These functions are solutions to the modified eigenvalue equation

$$-\frac{1}{\sqrt{\rho_0(\boldsymbol{r})}}\frac{\partial}{\partial r_j}\left(c_{ijkl}(\boldsymbol{r})\frac{\partial}{\partial r^l}\frac{f_k^{(n)}}{\sqrt{\rho_0(\boldsymbol{r})}}\right) = \Omega_n^2 f_i^{(n)}, \tag{12}$$

which is Hermitian with respect to the standard inner product known from electronic quantum mechanics. An alternative is to retain the original modes and add the offending density as a weight function to the inner product:

$$\langle \boldsymbol{u}^{(m)}|\boldsymbol{u}^{(n)}\rangle_\rho = \int \mathrm{d}r^2\, \rho_0(\boldsymbol{r})[\boldsymbol{u}^{(m)}(\boldsymbol{r})]^* \cdot \boldsymbol{u}^{(n)}(\boldsymbol{r}). \tag{13}$$

The mechanical states are orthogonal with respect to this form, the operator Eq. (10) is Hermitian and has the useful property that it provides the modal energy density (i.e., energy per unit waveguide length) as its norm:

$$\mathcal{E}_b = 2\Omega^2\langle \boldsymbol{u}|\boldsymbol{u}\rangle_\rho. \tag{14}$$

This normalization constant will be required in the derivation of the Hamiltonian framework. When dividing by an acoustic quantum of energy $\hbar\Omega$, it also allows to normalize states to the number of excitations as is typical in quantum formulations.

2.2 Essentials of linear and nonlinear electromagnetism

The foundations of waveguide optics have been exhaustively presented in any number of texts (Agrawal, 2012; Okamoto, 2006; Vassallo, 1991). Here we outline the minimum formalism required to define the problem precisely.

In SI units, Maxwell's equations in nonconducting media with no free charge or currents have the form

$$\nabla \cdot \boldsymbol{D} = 0, \qquad \nabla \times \boldsymbol{H} = \frac{\partial \boldsymbol{D}}{\partial t}, \tag{15a}$$

$$\nabla \cdot \boldsymbol{B} = 0, \qquad \nabla \times \boldsymbol{E} = -\frac{\partial \boldsymbol{B}}{\partial t}. \tag{15b}$$

Materials are assumed nonmagnetic ($\boldsymbol{B} = \mu_0 \boldsymbol{H}$) where μ_0 is the permeability of free space, and the linear dielectric response is represented by the *relative* dielectric tensor ϵ_{ij}:

$$D_i = \epsilon_0 E_i + P_i^{(\mathrm{L})}[\boldsymbol{E}] = \epsilon_0 \epsilon_{ij} E_j, \tag{16}$$

where $\boldsymbol{P}^{(\mathrm{L})}(\boldsymbol{r})$ is the linear polarization density.[c] We will have some use for the inverse relative dielectric tensor β_{ij}, defined as $\beta_{ij}\epsilon_{jk} = \delta_{ik}$ where δ_{ik} is the Kronecker delta symbol.

In many areas of nonlinear waveguide optics, the nonlinear response is incorporated by including a nonlinear polarization term $\boldsymbol{P}^{(\mathrm{NL})}$, which usually takes the form of a series expansion in the electric field $\boldsymbol{E}$ with nonlinear susceptibilities $\chi^{(2)}_{ijk}, \chi^{(3)}_{ijkl}$ (Boyd, 2008). (In the quantum formalism described below, a corresponding expansion in the displacement field $\boldsymbol{D}$ is actually more suitable (Quesada et al., 2021).)

In our approach to the Brillouin problem, the optomechanical coupling enters through an implicit dependence of the dielectric tensor $\epsilon_{ij}(\boldsymbol{r})$ on the strain field, and we will generally not need to refer to nonlinear susceptibilities. However, they can certainly be included in order to account for other nonlinearities which may be acting simultaneously, such as four-wave mixing and self- or cross-phase modulation.

2.2.1 Modal solutions

The formulation of the electromagnetic modal problem is very similar to the elastic case. Again, there is a vast literature on analytic and numerical techniques for solving both cavity and waveguide modal problems in the electromagnetic domain. Due to the utility of a Hermitian mode expansion, we will simply mention the useful "master equation"[d] approach that particularly came to prominence in the photonic crystal literature (Joannopoulos et al., 2011; Johnson and Joannopoulos, 2001). Defining modal expansions

[c] For compactness, we do not bother with a subscript r to indicate that ϵ_{ij} is the relative permittivity tensor. Factors of the permittivity of free space ϵ_0 are always explicitly included.

[d] Useful, but perhaps unfortunately named. As quantum treatments of Brillouin and optomechanical problems increasingly use open systems techniques involving *quantum master equations* for the evolution of the density operator, we suggest referring to this equation as simply the vector Helmholtz equation.

$$H(r,t) = h(r)e^{-i\omega t} + \text{c.c.}, \tag{17a}$$
$$E(r,t) = e(r)e^{-i\omega t} + \text{c.c.}, \tag{17b}$$
$$D(r,t) = d(r)e^{-i\omega t} + \text{c.c.}, \tag{17c}$$

one looks for eigensolutions to the vector Helmholtz equation for the magnetic field $\boldsymbol{H}$ in the form

$$\nabla \times \left[\overleftrightarrow{\beta}(r)\nabla \times H(r)\right] = \frac{\omega^2}{c^2} H(r), \tag{18}$$

which is manifestly Hermitian. Once $\boldsymbol{H}$ is known, the $\boldsymbol{D}$ and $\boldsymbol{E}$ fields follow directly from Eqs. (15a) and (16). As in the mechanical case, it may be useful to introduce a specialized inner product for the electric fields:

$$\langle D^{(m,k_m)} | D^{(n,k_n)} \rangle_\beta = \frac{1}{\epsilon_0} \int \mathrm{d}^2 r \, [D_i^{(m,k_m)}(r)]^* \beta_{ij}(r) D_j^{(n,k_n)}(r), \tag{19}$$

which as before provides a way to normalize, e.g., numerically obtained eigenmodes to a given unit of energy per unit length of waveguide:

$$\mathcal{E}_n = 2\langle D^{(n,k_n)} | D^{(n,k_n)} \rangle_\beta, \tag{20}$$

which we again will need in the following.

3. The Hamiltonian approach to Brillouin scattering

We are now ready to turn to the derivation of the CMEs that are the most common tool for modeling Brillouin scattering. Among several ways of deriving the CMEs, the Hamiltonian approach has two distinct advantages. First, it provides a natural connection between the classical theory of Brillouin and the quantum theory required for treatment of nonclassical statistics and quantum noise in Brillouin scattering. Second, it presents some necessary approximations in a straightforward way and provides a way to ensure that they are applied consistently, i.e., in a way that does not jeopardize the conservation of important quantities such as energy. In this section, we draw closely from the formalism outlined in three references (Sipe and Steel, 2016; Wolff et al., 2015b, 2017).

3.1 The opto-acoustic Hamiltonian

The starting point of any Hamiltonian treatment is the Hamiltonian itself, i.e., an expression for the total energy expressed in terms of canonical variables. In any optomechanical problem, this includes (at least) the total mechanical energy and the total electromagnetic energy in the system[e] :

$$\mathcal{H} = \mathcal{H}^{(\mathrm{mech})} + \mathcal{H}^{(\mathrm{EM})}. \tag{21}$$

First, we discuss the mechanical Hamiltonian. As mentioned in Section 2, the deformation of a continuous body can be described either in terms of a displacement field $\boldsymbol{U}$ and associated force densities, or in terms of a strain $\overleftrightarrow{S}$ and an associated stress field $\overleftrightarrow{T}$. We choose the latter so that the mechanical Hamiltonian takes the form

$$\mathcal{H}^{(\mathrm{mech})} = \frac{1}{2} \int \mathrm{d}^3 r \left[\frac{|\boldsymbol{\pi}(\boldsymbol{r})|^2}{\rho(\boldsymbol{r})} + S_{ij}(\boldsymbol{r}) c_{ijkl}(\boldsymbol{r}) S_{kl}(\boldsymbol{r}) \right], \tag{22}$$

where the integration is carried out over all of space and we have again employed Einstein's summation convention to sum over the indices i, j, k, and l in the second term.

In Eq. (22), the displacement $\boldsymbol{U}(\boldsymbol{r}, t)$ (which enters implicitly through $\overleftrightarrow{S}$) and the momentum density

$$\boldsymbol{\pi}(\boldsymbol{r}) = \rho(\boldsymbol{r}) \frac{\partial \boldsymbol{U}}{\partial t}, \tag{23}$$

are canonically conjugate quantities, which can be treated as either classical fields or quantum field operators as one prefers. On the other hand, the density $\rho(\boldsymbol{r})$ and stiffness $c_{ijkl}(\boldsymbol{r})$ are not dynamical variables and are treated as parameters that completely define the mechanical properties.

Depending on one's preference for a classical or quantum[f] picture, both the momentum continuity relation

$$\frac{\partial}{\partial t} \pi_j(\boldsymbol{r}) + \frac{\partial}{\partial r_i} T_{ij}(\boldsymbol{r}) = 0, \tag{24}$$

and the wave Eq. (5) (without the viscous damping term) can be recovered from (22) by setting the Poisson brackets $\{U_i(\boldsymbol{r}), \pi_j(\boldsymbol{r}')\} = \delta_{ij}\, \delta(\boldsymbol{r} - \boldsymbol{r}')$ and using Hamilton's equations, or by choosing the commutators

$$[U_i(\boldsymbol{r}), \pi_j(\boldsymbol{r}')] = \mathrm{i}\hbar \delta_{ij} \delta(\boldsymbol{r} - \boldsymbol{r}'), \tag{25}$$

[e] Typically we might also expect a coupling or interaction term. In the development that follows, the interaction will arise from the strain-dependence of $\mathcal{H}^{(\mathrm{EM})}$.

[f] In general, we will denote quantum operators without hats except where necessary to avoid ambiguity.

and evaluating the Heisenberg equation for the evolution of each operator $\hat{O}$ (see Section 3.3):

$$\frac{\partial}{\partial t}\hat{O} = \frac{i}{\hbar}[\mathcal{H}, \hat{O}]. \tag{26}$$

Thus, for example, Eq. (2) in the absence of external driving forces is recovered from the Heisenberg equations:

$$\frac{\partial U_i(\boldsymbol{r})}{\partial t} = \frac{i}{\hbar}[\mathcal{H}^{(\mathrm{mech})}, U_i] \tag{27a}$$

$$= \frac{i}{2\hbar}\int d^3r' \underbrace{\left[\frac{|\pi(\boldsymbol{r}')|^2}{\rho(\boldsymbol{r}')}, U_i(\boldsymbol{r})\right]}_{=\rho^{-1}\sum_j [U_i, \pi_j^2]} + \underbrace{[U_i, \overleftrightarrow{S} : \overleftrightarrow{c} : \overleftrightarrow{S}]}_{=0}$$

$$= \frac{i}{2\hbar}\sum_j \int d^3r' \frac{1}{\rho(\boldsymbol{r}')}[\pi_j, U_i]\pi_j + \pi_j[\pi_j, U_i] \tag{27b}$$

$$= \frac{i}{2\hbar}\sum_j \int d^3r' \frac{-2i\hbar}{\rho(\boldsymbol{r}')}\pi_j\delta_{ij}\delta(\boldsymbol{r} - \boldsymbol{r}')$$

$$= \frac{1}{\rho(\boldsymbol{r})}\pi_i(\boldsymbol{r}),$$

$$\frac{\partial \pi_i(\boldsymbol{r})}{\partial t} = \frac{i}{\hbar}[\mathcal{H}^{(\mathrm{mech})}, \pi_i], \tag{27c}$$

where we added the evaluation of the first commutator for illustration purposes.

A feature of mechanical systems with finite deformations is that they are inherently nonlinear. Even assuming only materials that satisfy Hooke's law, i.e., whose bulk response to deformation is perfectly linear, any system other than an infinitely extended homogeneous block exhibits mechanical nonlinearity. The reason is that the deformation of the body changes its geometry and thereby its resistance to further deformation. An extreme case of this is buckling (Landau and Lifshitz, 1986). Our mechanical Hamiltonian equation (22) neglects any such nonlinear effect (as well as mechanical bulk nonlinearities, for that matter). The justification is that the relative change to the "deformation resistance" of a waveguide can be neglected to first approximation, because of the small strains that typically occur in Brillouin scattering and because of the dominant linear response.

The second term in Eq. (21) is the electromagnetic Hamiltonian

$$\mathcal{H}^{(\mathrm{EM})} = \frac{1}{2\epsilon_0}\int \mathrm{d}^3r\, D_i(\boldsymbol{r})\beta_{ij}(\boldsymbol{r})D_j(\boldsymbol{r}) + \frac{1}{2\mu_0}\int \mathrm{d}^3r\, |\boldsymbol{B}(\boldsymbol{r})|^2, \tag{28}$$

where, again, the integral covers all of space. The choice of the $\boldsymbol{D}$ and $\boldsymbol{B}$ fields as the fundamental degrees of freedom can be unfamiliar on a first encounter, but it has a long heritage in medium quantization (Born and Infeld, 1934; Sipe and Steel, 2016) and it has recently been recognized that it is essential for ensuring a physically consistent quantization in the nonlinear regime (Quesada and Sipe, 2017).

As with the mechanical problem, the electromagnetic fields depend on the boundary conditions imposed by the distribution of materials but the presence of a mechanical deformation field modifies the electromagnetic problem. A simple illustration of this follows from the fact that the dispersion relation of say, a silicon nanowire, depends on the height and width of the nanowire. Thus, if the wire's geometry changes, so does the dispersion relation. In this context it does not matter if the deformation is static (e.g., due to fabrication tolerances) or dynamic as long as the deformation is slow compared to one optical cycle. Thus, the motion of material boundaries constitutes an effective optomechanical nonlinearity. In addition to this, the bulk permittivity of a material is also directly modified by the presence of a strain field S_{ij} via the photoelastic effect according to

$$\Delta\beta_{ij}(\boldsymbol{r};\ \overleftrightarrow{S}) \equiv [\Delta\beta(\boldsymbol{r};\ \overleftrightarrow{S})]_{ij} = p_{ijkl}(\boldsymbol{r})S_{kl}(\boldsymbol{r}), \tag{29}$$

where p_{ijkl} is the photoelastic Pockels tensor (Nelson and Lax, 1970). In the traditional literature on SBS in bulk and optical fibers, this has been the dominant effect. The moving boundary effect is arguably more fundamental and in contemporary submicron waveguides it can be the dominant effect. It is also the primary origin of the optomechanical effects to be discussed in chapter "The Convergence of cavity optomechanics and Brillouin scattering" by Schmidt et al.

In the case of an anisotropic dielectric, there is also the related roto-optic effect based on the local rotation (or twist) field $R_{ij} = \frac{1}{2}\left(\frac{\partial u_i}{\partial r_j} - \frac{\partial u_j}{\partial r_i}\right)$ (Nelson and Lax, 1970) according to

$$\Delta\beta_{ij}(\boldsymbol{r};\ \overleftrightarrow{R}) = r_{ijkl}(\boldsymbol{r})R_{kl}(\boldsymbol{r}), \tag{30}$$

with the roto-optic tensor r_{ijkl}, which takes the form (Smith et al., 2017):

$$r_{ijkl} = \frac{1}{2}\left(\beta_{il}\delta_{jk} + \beta_{lj}\delta_{ik} - \beta_{ik}\delta_{jl} - \beta_{kj}\delta_{il}\right). \tag{31}$$

The main point is that through all three mechanisms, the permittivity distribution $\epsilon_{ij}(\boldsymbol{r})$ explicitly depends on the strain field S_{ij}. Since the material distribution determines the electric and magnetic field distributions, this means that they, too, depend on the strain field, although indirectly through $\epsilon_{ij}(\boldsymbol{r})$, and the electromagnetic Hamiltonian is better written as

$$\begin{aligned}&\mathcal{H}^{(\mathrm{EM})}[\boldsymbol{B},\boldsymbol{D};\ \overleftrightarrow{S}]\\&=\frac{1}{2\epsilon_0}\int \mathrm{d}^3r\ D_i(\boldsymbol{r};\ \overleftrightarrow{S})\beta_{ij}(\boldsymbol{r};\ \overleftrightarrow{S})D_j(\boldsymbol{r};\ \overleftrightarrow{S})+\frac{1}{2\mu_0}\int \mathrm{d}^3r\ |\boldsymbol{B}(\boldsymbol{r};\ \overleftrightarrow{S})|^2.\end{aligned} \tag{32}$$

The dependence of these terms on the strain $\overleftrightarrow{S}$ is highly nonlinear (especially near material interfaces). Luckily, Brillouin scattering is typically associated with weak deformations and can be treated perturbatively.[g] This splits the total electromagnetic Hamiltonian into that of the undeformed system and the optomechanical interaction Hamiltonian:

$$\mathcal{H}^{(\mathrm{EM})}[\boldsymbol{B},\boldsymbol{D};\ \overleftrightarrow{S},\ \overleftrightarrow{R},\boldsymbol{U},\boldsymbol{V}]\approx\mathcal{H}_0^{(\mathrm{EM})}[\boldsymbol{B},\boldsymbol{D}]+\mathcal{H}^{(\mathrm{OM})}[\boldsymbol{B},\boldsymbol{D};\ \overleftrightarrow{S},\ \overleftrightarrow{R},\boldsymbol{U},\boldsymbol{V}]. \tag{33}$$

The linear electromagnetic Hamiltonian $\mathcal{H}_0^{(\mathrm{EM})}$ is the same as in Eq. (28) for the undeformed system. The derivation of the interaction term is lengthy and requires careful attention to the electromagnetic continuity conditions at dielectric interfaces (Johnson et al., 2002). For details we refer the reader to the literature (Rakich et al., 2012; Sipe and Steel, 2016; Wolff et al., 2015b) and only state the result. There are four interaction terms

$$\mathcal{H}^{(\mathrm{OM})}=\underbrace{\mathcal{H}^{(\mathrm{MB})}[\boldsymbol{U}]}_{\text{moving-boundary}}+\underbrace{\mathcal{H}^{(\mathrm{PE})}[\overleftrightarrow{S}]}_{\text{photoelastic}}+\underbrace{\mathcal{H}^{(\mathrm{RO})}[\overleftrightarrow{R}]}_{\text{roto-optic}}+\underbrace{\mathcal{H}^{(\mathrm{MP})}[\boldsymbol{V}]}_{\text{moving-polarization}}, \tag{34}$$

where we have omitted the dependence on the electromagnetic field $\boldsymbol{B}$ and $\boldsymbol{D}$ for the sake of brevity. The first term (moving-boundary) depends on the local displacement vector $\boldsymbol{U}$ at a material interface, the second (photoelastic) depends on the strain field $\overleftrightarrow{S}$ in any bulk dielectric, the third (roto-optic) depends on the local rotation $\overleftrightarrow{R}$ or twisting and only contributes in materials with natural or artifical optical anisotropy, and the last (moving-polarization) depends on the local velocity vector in any bulk dielectric, but is very weak

[g] In principle, there is a corresponding dependence of the elastic stiffness c_{ijkl} on the electric displacement field $\boldsymbol{D}$ but this is weaker by the ratio of elastic and optical frequencies $\Omega/\omega\approx 10^{-5}$ (Sipe and Steel, 2016).

and can usually be neglected (Wolff et al., 2015b). The three relevant terms take the forms:

$$\mathcal{H}^{(\mathrm{PE})} = \frac{1}{2\epsilon_0}\int \mathrm{d}^3r\ D_i(\boldsymbol{r})D_j(\boldsymbol{r})p_{ijkl}(\boldsymbol{r})S_{kl}(\boldsymbol{r}); \tag{35a}$$

$$\mathcal{H}^{(\mathrm{RO})} = \frac{1}{2\epsilon_0}\int \mathrm{d}^3r\ D_i(\boldsymbol{r})D_j(\boldsymbol{r})r_{ijkl}(\boldsymbol{r})R_{kl}(\boldsymbol{r}); \tag{35b}$$

$$\begin{aligned}\mathcal{H}^{(\mathrm{MB})} = &\frac{1}{2}\int_C \mathrm{d}^2r(\hat{n}\cdot\boldsymbol{U})\{(\epsilon_a-\epsilon_b)\epsilon_0[\hat{n}\times\boldsymbol{E}(\boldsymbol{r})]\cdot[\hat{n}\times\boldsymbol{E}(\boldsymbol{r})]\\ &-(\epsilon_a^{-1}-\epsilon_b^{-1})\epsilon_0^{-1}[\hat{n}\cdot\boldsymbol{D}(\boldsymbol{r})][\hat{n}\cdot\boldsymbol{D}(\boldsymbol{r})]\},\end{aligned} \tag{35c}$$

where for the moving-boundary part we assume two homogeneous materials with ϵ_a and ϵ_b separated by a contour C with local normal vector $\hat{n}$ (pointing from ϵ_b to ϵ_a) (Wolff et al., 2021). As we will further discuss in Section 3.4, the moving-boundary term also captures radiation pressure, the photoelastic term covers electrostriction and the roto-optic term describes the coupling between the optically induced torque field and local rotation (or twist) of the acoustic mode. In high-index nanophotonic waveguides, it is usually both sufficient but also crucial to include the moving-boundary (radiation pressure) and the photoelastic (electrostrictive) terms (Rakich et al., 2012).

One fundamental shortcoming of the classical Hamiltonian treatment is that the incorporation of dissipation from first principles is less convenient than in other approaches. In quantum-mechanical formulations, this is typically accomplished by introducing a bath of environment oscillators which are then traced over, leading to either master equation or quantum Langevin equation treatments (Bowen and Milburn, 2015; Breuer and Petruccione, 2002). However, we skip these complexities at this point. Instead, we adopt the more pragmatic stance of deriving the equations of motion from the Hamiltonian and then phenomenologically introducing classical loss after the fact.

3.2 Eigenmode expansion

Based on the total Hamiltonian, we could now proceed to derive equations of motion, but we would end up in an "ab-initio"-like formulation, which would be impractical as discussed in the first section. Therefore, we insert a few approximations at this point. This eventually will lead us to the coupled mode picture that is the most common way to model Brillouin scattering. First, we expand both the optical and the mechanical fields into sets of

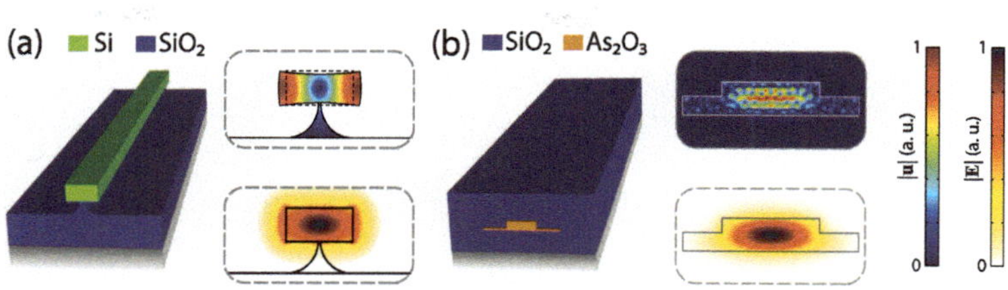

Fig. 1 Optical and mechanical modes involved in Brillouin scattering. (A) Schematic of a silicon waveguide (*left*) and its mechanical (*upper*) and optical (*lower*) mode spatial profiles. This example resembles the waveguide explored by Laer et al. (2015), where forward Brillouin scattering was investigated. (B) Schematic of a As_2S_3 waveguide and its mechanical and optical mode spatial profiles, resembling the waveguide explored by Pant et al. (2011), where backward Brillouin scattering at 11 GHz was observed.

orthogonal eigenmodes. Within a quantum-mechanical treatment this is by far the most common approach[h] in order to quantize the fields, but it is also highly advantageous within a classical formalism. In the following we will present the main steps in the classical and the quantum-mechanical picture side by side.

We assume that the Brillouin scattering takes place in a waveguide, which for convenience we assume to be aligned parallel to the z-axis. A typical configuration is illustrated in Fig. 1. Next, we pick an expansion basis for each of the mechanical and the electromagnetic fields. For this, we choose the eigenmodes of the unperturbed mechanical and optical fields, i.e., in the quantum-mechanical case the eigenstates of the respective Hamiltonians in Eqs. (22) and (28). In the classical case, the eigenmodes can be found by solving the lossless mechanical and optical wave equations, e.g., via a finite-element solver (Sturmberg et al., 2018). Due to the lack of dissipation and assuming bound states, these eigenmodes form complete sets of functions and can be used to expand the physical fields. Since the system is a waveguide, the states are identified by their longitudinal wave numbers (k for optical and q for mechanical modes) and optional discrete band indices n. The expansions for the electromagnetic and mechanical fields are all quite similar; for the electric field $\boldsymbol{E}$ and the mechanical strain field $\overleftrightarrow{S}$ they read

[h] We note that there are alternatives to the ubiquitous second quantization in terms of classical eigenmodes. For example, it is possible to formulate a quantum theory on the grounds of canonical commutator relations such as Eq. (25) or $[\boldsymbol{D}(\boldsymbol{r}, t), \boldsymbol{B}(\boldsymbol{r}, t)]$ alone. Furthermore, it is possible (and in open and lossy systems potentially advantageous) to formulate a quantum theory based on propagators derived from classical Green functions (Vogel and Welsch, 2006).

$$\boldsymbol{E}(\boldsymbol{r},t) = \sum_n \int_{-\infty}^{\infty} \mathrm{d}k\ \tilde{a}_{nk}(t) \exp(\mathrm{i}kz - \mathrm{i}\omega_{nk}t)\boldsymbol{e}^{(nk)}(x,y)\ + \text{h.c.}, \tag{36a}$$

$$S_{ij}(\boldsymbol{r},t) = \int_{-\infty}^{\infty} \mathrm{d}q\ \tilde{b}_q(t) \exp(\mathrm{i}qz - \mathrm{i}\Omega_q t) s_{ij}^{(q)}(x,y)\ + \text{h.c.}, \tag{36b}$$

where $\boldsymbol{e}^{(nk)}(x, y)$ is the electric field of the appropriately normalized (see below) optical eigenmode with band index n and wave number k, $s_{ij}^{(q)}(x,y)$ is likewise the strain field of a mechanical eigenmode, and ω_{nk} and Ω_q are the (angular) optical and mechanical eigenfrequencies, respectively. Finally, $\tilde{a}_{nk}$ and $\tilde{b}_q$ are complex-valued expansion coefficients in the classical case and annihilation operators in the quantum-mechanical case. In the same spirit, we have used h.c. to denote the Hermitian conjugate,[i] which in the classical case should be read as the complex conjugate (c.c.). We will continue with this convention for the remainder of the chapter.

We should note that technically the mode expansion of the mechanical field should also include a summation over all bands in order to have a complete basis. However, we neglect this, because in the end we only consider one acoustic mode at a time and it would clutter the notation unnecessarily.

The resulting quantum-mechanical Hamiltonians take the very simple forms

$$\mathcal{H}^{(\mathrm{EM})} = \sum_n \int_{-\infty}^{\infty} \mathrm{d}k\ \hbar\omega_{nk}\ \tilde{a}_{nk}^{\dagger}(t)\tilde{a}_{nk}(t), \tag{37a}$$

$$\mathcal{H}^{(\mathrm{mech})} = \int_{-\infty}^{\infty} \mathrm{d}q\ \hbar\Omega_q\ \tilde{b}_q^{\dagger}(t)\tilde{b}_q(t). \tag{37b}$$

Their classical counterparts are identical except that the Hermitian conjugate is replaced with a complex conjugation and the energy quanta $\hbar\omega_{nk}$ and $\hbar\Omega_q$ with classical mode normalization constants. We defer discussion of the proper choice of these constants for now until the end of this section. In order for the expressions in Eq. (37) to have dimensions of energy, the k-space amplitudes a_{nk} and b_q must have dimensions of square root of length.

[i] We note that the concept of a Hermitian conjugate may be ambiguous in a problem that involves both physical quantities expressed as second-rank complex tensors and quantum-mechanical operators. To be clear: within this chapter the Hermitian conjugate performs complex conjugations on all non-operator quantities and replaces quantum-mechanical operators with their adjoints with respect to the QM Hilbert space. It does *not* transpose classical second-rank tensors.

The expansions Eq. (36) have the disadvantage that the fields are expressed as "envelopes" $\tilde{a}_{nk}$ and $\tilde{b}_q$, that "live" in k-space rather than in real space. Therefore, we apply a Fourier transform

$$a_n(z,t) = (2\pi L)^{-1/2} \int_{-\infty}^{\infty} \mathrm{d}k \ \exp[\mathrm{i}(k-k_n)z - \mathrm{i}\omega_n t]\tilde{a}_{nk}(t), \tag{38a}$$

$$b(z,t) = (2\pi L)^{-1/2} \int_{-\infty}^{\infty} \mathrm{d}q \ \exp[\mathrm{i}(q-q_0)z - \mathrm{i}\Omega_0 t]\tilde{b}_q(t), \tag{38b}$$

where we have chosen "center" wave numbers k_n with associated eigenfrequency ω_n for each optical mode and similarly for the mechanical mode. We have introduced an additional constant L that would be missing in most dedicated quantum-mechanical treatments. It has units of length and its meaning is that of a *quantization length*, i.e., the length over which an optical or acoustic quantum is "spread out." Typical choices could be the length of the waveguide or simply 1 m. We have introduced this factor in order to obtain envelope operators $a_n(z, t)$ and $b(z, t)$ that are dimensionless like their classical counterparts. Without this factor, the envelopes end up with dimensions of $\mathrm{m}^{-1/2}$, as is common in quantum treatments of waveguides (Sipe and Steel, 2016). A particular advantage of our choice here is that all mode-related quantities retain "intuitive" units. The electric modal field $\boldsymbol{e}^{(nk_n)}$ profile, for example, has the expected units of V/m that one would find in solutions from a finite-element solver, rather than units of $\mathrm{V}/\sqrt{\mathrm{m}}$ which would result without the factor $1\sqrt{L}$. Furthermore, derived quantities (such as the overlap products that will be introduced in Eq. (49)), carry over more easily between the quantum theory for SBS in waveguides and its classical and cavity-bound counterparts (see Section 5). As in other branches of physics that invoke "box normalization," for all calculations of measurable quantities, the final result will be found not to depend on L. For further discussion, please refer to Section 3.4.1.

We now have obtained "proper" envelope functions, i.e., functions that describe the amplitudes of the physical fields in space and time. However, since the mode patterns ($\boldsymbol{d}(x, y)$, $\boldsymbol{b}(x, y)$ etc.) depend to some degree on the wave number, the expansions Eq. (37) cannot exactly be translated to a product between the envelope and the eigenmode. This is where the "center" wave numbers come into play; assuming that the k-space occupied by the pulse is small enough that the mode patterns are approximately constant, we may approximate:

$$\boldsymbol{E}(\boldsymbol{r},t) = \sum_n a_n(z,t)\boldsymbol{e}^{(nk_n)}(x,y) + \text{h.c.}; \tag{39a}$$

$$S_{ij}(\boldsymbol{r},t) = b(z,t)s_{ij}^{(q_0)}(x,y) + \text{h.c.}, \tag{39b}$$

where the real-space envelope mode operators $a_n(z, t)$ and $b(z, t)$ are dimensionless and obey equal-time commutation relations

$$\left[a_m(z,t), a_n^\dagger(z',t)\right] = L\delta_{mn}\delta(z-z'), \tag{40a}$$

and

$$\left[b(z,t), b^\dagger(z',t)\right] = L\delta(z-z'), \tag{40b}$$

provided the optical and acoustic modes have been normalized (see Eqs. 14 and 20) according to

$$\mathcal{E}_n^{(\text{QM})} = \frac{\hbar\omega_n}{L}; \qquad \mathcal{E}_b^{(\text{QM})} = \frac{\hbar\Omega_0}{L}. \tag{41}$$

All other commutators vanish. This constitutes a slowly varying envelope approximation (SVEA), which is the first approximation required in the derivation of the equations of motion.

We pay for the real-space representation of the expansions Eqs. (39) by slightly more complicated expressions for the Hamiltonians. To be precise, we need to translate the nontrivial dispersion relations represented by $\omega_n(k)$ and $\Omega(q)$ to real space. To this end, we Taylor expand the dispersion relations with respect to k and q around k_n and q_0 and then use the Fourier "translation rules"

$$\mathrm{i}k \leftrightarrow \partial_z, \quad \text{or} \quad \mathrm{i}q \leftrightarrow \partial_z, \quad \text{respectively.} \tag{42}$$

Furthermore, we note that we already included the linear carrier terms $\mathrm{i}\omega_n$, $\mathrm{i}\Omega_0$, $\mathrm{i}k_n$ and $\mathrm{i}q_0$ in the definitions of the real-space amplitude operators (Eq. 38), so we obtain:

$$\mathcal{H}^{(\text{EM})} = -\frac{1}{L}\sum_n \int_0^L \mathrm{d}z\, \hbar[\mathrm{i}v_n\partial_z + \cdots]a_n(z,t)^\dagger a_n(z,t), \tag{43a}$$

$$\mathcal{H}^{(\text{mech})} = -\frac{1}{L}\int_0^L \mathrm{d}z\, \hbar[\mathrm{i}v_b\partial_z + \cdots]b(z,t)^\dagger b(z,t), \tag{43b}$$

where $v_n = \partial\omega_n(k)/\partial k$ and $v_b = \partial\Omega(q)/\partial q$ are the group velocities of the various optical and acoustic modes, respectively. The expansion of the dispersion relations is basically a sequel to the SVEA in Eqs. (39). A more detailed description of the procedure including the incorporation of higher order dispersion terms can be found in Sipe and Steel (2016) and Wolff et al. (2017).

In the quantum-mechanical treatment, the envelopes $a_n(z, t)$ and $b(z, t)$ are operators that describe the annihilation of a photon or phonon in mode n at time t and position z along the waveguide. We can now substitute the expressions Eqs. (39) into the optomechanical interaction Hamiltonians in Eqs. (35). We will discuss this for the photoelastic term Eq. (35a); the steps for the other interaction terms are analogous, if sometimes more involved, and can be found discussed in the literature (Sipe and Steel, 2016; Wolff et al., 2015b). The photoelastic Hamiltonian (35a) is the integral over the interaction energy density

$$\mathcal{I}^{(\mathrm{PE})} = \frac{1}{2\epsilon_0} D_i(\mathbf{r}) D_j(\mathbf{r}) p_{ijkl}(\mathbf{r}) S_{kl}(\mathbf{r}). \tag{44}$$

Substituting Eqs. (39) leads to

$$\begin{aligned}\mathcal{I}^{(\mathrm{PE})} = \frac{1}{2\epsilon_0} p_{ijkl}(\mathbf{r}) \sum_{nn'} &[a_n(z,t) d_i^{(nk_n)}(x,y) + \mathrm{h.c.}] \\ &\times [a_{n'}(z,t) d_j^{(n'k_{n'})}(x,y) + \mathrm{h.c.}][b(z,t) s_{kl}^{(q_0)}(x,y) + \mathrm{h.c.}].\end{aligned} \tag{45}$$

When we now expand the product of the three brackets, we find terms such as $a_n(z, t)a_{n'}(z, t)$, which correspond to oscillations at the frequency $\omega_n + \omega_{n'} \approx 2\omega_n$. These terms must be neglected, firstly because they are too fast for the mechanical system to follow, but also, because they contradict our earlier assumption that the amplitudes $a_n(z, t)$ are slowly varying envelopes. This is a common feature in light–matter interactions and those terms are usually dropped in a so-called rotating wave approximation (RWA). In the theory of Brillouin scattering, this comes in two stages. The first one is to drop terms like $a_n a_{n'}$ or $a_n^\dagger a_{n'}^\dagger$ and only retain $a_n a_{n'}^\dagger$ and $a_n^\dagger a_{n'}$, which oscillate at the difference frequency $\omega_n - \omega_{n'}$, which is assumed to be on the order of the mechanical frequency Ω. Using the symmetry $p_{ijkl} = p_{jikl}$ of the Pockels tensor, this results in the expression:

$$\begin{aligned}\mathcal{I}^{(\mathrm{PE})} = \frac{p_{ijkl}(\mathbf{r})}{2\epsilon_0} \sum_{nn'} &\left(a_n^\dagger(z,t) a_{n'}(z,t) [d_i^{(nk_n)}(x,y)]^* d_j^{(n'k_{n'})}(x,y) + \mathrm{h.c.} \right) \\ &\times \left(b(z,t) s_{kl}^{(q_0)}(x,y) + \mathrm{h.c.} \right).\end{aligned} \tag{46}$$

The same reasoning also applies to the product of the form $(a_n^\dagger a_{n'} + a_n a_{n'}^\dagger)(b + b^\dagger)$. The product has in total four terms, two of which oscillate at twice the mechanical frequency and the other two at a much slower frequency. The former terms are therefore commonly neglected as part of a restrictive SVEA as discussed in the following. This is necessary if one aims to represent different Stokes orders by individual envelopes. However, there are cases where this is not desirable or possible (Piotrowski et al., 2021; Velchev et al., 1999; Wolff et al., 2017; Yuan et al., 2019).

Here, we present the case of the restrictive RWA. The summations in Eq. (46) cover all optical band indices for n and n' independently, so we actually sum each possible combination $n \neq n'$ twice. In the case $n \neq n'$, the creation and annihilation operators commute and we include the same term twice for each combination—once with $\omega_n > \omega_{n'}$ and once with $\omega_n < \omega_{n'}$. This means we can account for this double counting by adding a factor 2 and only retaining one of them. In the case $n = n'$, the optical operators do not commute: $a_n(z)a_n^\dagger(z') = a_n^\dagger(z')a_n(z) + L\delta(z - z')$ (see Eq. 40a), so in addition to the factor two there is a further contribution. This scales with $\hbar\omega/2$ but is independent of the photon number, and thus describes static background stresses and forces acting on the waveguide due to photon vacuum fluctuations. The physical interpretation is that (if integrated over all photon energies and mode combinations) this nontrivial commutator constitutes the Casimir-effect (for the moving-boundary term) and similar dispersion forces (Wolff et al., 2017). They are of fundamental interest, but because of their static nature do not contribute to the dynamics of optomechanical interactions and can be ignored in the context of Brillouin scattering. We find the simplified interaction integrand:

$$\begin{aligned} \mathcal{I}^{(\mathrm{PE})} = \frac{1}{\epsilon_0} \sum_{nn'}^{\omega_n \geq \omega_{n'}} & a_n^\dagger(z,t) a_{n'}(z,t) b(z,t) \\ & \times \left[d_i^{(nk_n)}(x,y)\right]^* d_j^{(n'k_{n'})}(x,y) p_{ijkl}(\boldsymbol{r}) s_{kl}^{(q_0)}(x,y) + \text{h.c.}, \end{aligned} \tag{47}$$

where the expression above the summation indicates that only combinations (n, n') that satisfy this inequality should be included. If we now integrate this over all space to obtain the full photoelastic interaction Hamiltonian, we find that we can separate the envelopes from the mode profiles:

$$\mathcal{H}^{(\mathrm{PE})} = \int_0^L \mathrm{d}z \sum_{nn'}^{\omega_n \geq \omega_{n'}} Q_{nn'}^{(\mathrm{PE})} a_n^\dagger(z,t) a_{n'}(z,t) b(z,t) + \text{h.c.}; \tag{48a}$$

where the coupling

$$Q_{nn'}^{(\mathrm{PE})} = \frac{1}{\epsilon_0} \int \mathrm{d}^2 r \, [d_i^{(nk_n)}(x, y)]^* \, d_j^{(n'k_{n'})}(x, y) \, p_{ijkl}(x, y) \, s_{kl}^{(q_0)}(x, y), \quad (48b)$$

has dimensions of a linear energy density (i.e., energy per length). Likewise, we find expressions for the moving-boundary and roto-optic terms and can express the total optomechanical interaction Hamiltonian as:

$$\mathcal{H}^{(\mathrm{OM})} = \int_0^L \mathrm{d}z \sum_{nn'}^{\omega_n \geq \omega_{n'}} Q_{nn'}^{(\mathrm{tot})} a_n^\dagger(z, t) a_{n'}(z, t) b(z, t) + \mathrm{h.c.}; \quad (49a)$$

$$Q_{nn'}^{(\mathrm{tot})} = Q_{nn'}^{(\mathrm{MB})} + Q_{nn'}^{(\mathrm{PE})} + Q_{nn'}^{(\mathrm{RO})}, \quad (49b)$$

where the three couplings are

$$\begin{aligned} Q_{nn'}^{(\mathrm{MB})} &= \int_C \mathrm{d}^2 r \, [\hat{n} \cdot \boldsymbol{u}^{(q_0)}(x, y)] \\ &\quad \times \{ (\epsilon_a - \epsilon_b) \epsilon_0 [\hat{n} \times \boldsymbol{e}^{(nk_n)}(x, y)]^* \cdot [\hat{n} \times \boldsymbol{e}^{(n'k_{n'})}(x, y)] \\ &\quad - (\epsilon_a^{-1} - \epsilon_b^{-1}) \epsilon_0^{-1} [\hat{n} \cdot \boldsymbol{d}^{(nk_n)}(x, y)]^* [\hat{n} \cdot \boldsymbol{d}^{(n'k_{n'})}(x, y)] \}, \end{aligned} \quad (50a)$$

$$Q_{nn'}^{(\mathrm{PE})} = \frac{1}{\epsilon_0} \int \mathrm{d}^2 r \, [d_i^{(nk_n)}(x, y)]^* \, d_j^{(n'k_{n'})}(x, y) \, p_{ijkl}(x, y) \, s_{kl}^{(q_0)}(x, y), \quad (50b)$$

$$Q_{nn'}^{(\mathrm{RO})} = \frac{1}{\epsilon_0} \int \mathrm{d}^2 r \, [d_i^{(nk_n)}(x, y)]^* \, d_j^{(n'k_{n'})}(x, y) \, r_{ijkl}(x, y) \, r_{kl}^{(q_0)}(x, y). \quad (50c)$$

In these expressions, $r_{kl}^{(q_0)}$ is the local rotation matrix associated with the mechanical eigenmode, and the moving-boundary term is shown under the same assumptions as in Eqs. (35c).

3.3 Equations of motion

We are now ready to derive the equations of motion by applying the bosonic commutator relations (40) to the Hamiltonian. We find

$$-\mathrm{i}\hbar \partial_t a_n(z) = [\mathcal{H}^{(\mathrm{EM})}, a_n(z)] + [\mathcal{H}^{(\mathrm{OM})}, a_n(z)], \quad (51a)$$

$$-\mathrm{i}\hbar \partial_t b(z) = [\mathcal{H}^{(\mathrm{mech})}, b(z)] + [\mathcal{H}^{(\mathrm{OM})}, b(z)]. \quad (51b)$$

We will discuss the first of these equations and only state the result for the second one as its derivation is analogous. For the first commutator we insert the Taylor-approximation to the EM-Hamiltonian from Eq. (43a):

$$
\begin{aligned}
[\mathcal{H}^{(\mathrm{EM})}, a_n(z)] &\approx -\frac{1}{L}\sum_m \int_0^L \mathrm{d}z'\ \mathrm{i}\hbar v_m \partial_z a_m(z') \underbrace{[a_m(z)^\dagger, a_n(z')]}_{=-L\delta_{nm}\delta(z-z')} \\
&= \mathrm{i}\hbar v_n \partial_z a_n(z).
\end{aligned} \tag{52}
$$

Likewise we can evaluate the second commutator:

$$
\begin{aligned}
&[\mathcal{H}^{(\mathrm{OM})}, a_n(z)] \\
&= \sum_{mm'}^{\omega_m \geq \omega_{m'}} \int_0^L \mathrm{d}z' \left\{ Q_{mm'}^{(\mathrm{tot})} a_{m'}(z')[a_m^\dagger(z), a_n(z')] b(z') \right. \\
&\quad \left. + Q_{mm'}^{(\mathrm{tot})\ *} a_m(z')[a_n^\dagger(z), a_{m'}(z')] b^\dagger(z') \right\}
\end{aligned} \tag{53a}
$$

$$
= -\sum_{m'}^{\omega_n \geq \omega_{m'}} L Q_{nm'}^{(\mathrm{tot})} a_{m'}(z) b(z) - \sum_{m}^{\omega_m \geq \omega_n} L Q_{mn}^{(\mathrm{tot})*} a_m(z) b^\dagger(z). \tag{53b}
$$

These terms together provide the optical equations of motion; the acoustic equation of motion is found similarly. Assuming the optical and acoustic eigenmodes to be normalized according to Eq. (41), they read

$$
(\partial_t + v_n \partial_z) a_n = \sum_{m'}^{\omega_n \geq \omega_{m'}} -\mathrm{i}\frac{\omega_n Q_{nm'}^{(\mathrm{tot})}}{\mathcal{E}_n^{(\mathrm{QM})}} a_{m'} b - \mathrm{i}\sum_{m}^{\omega_m \geq \omega_n} \frac{Q_{mn}^{(\mathrm{tot})*}}{\mathcal{E}_n^{(\mathrm{QM})}} a_m b^\dagger, \tag{54a}
$$

$$
(\partial_t + v_b \partial_z) b = \sum_{mm'}^{\omega_m \geq \omega_{m'}} \frac{Q_{mm'}^{(\mathrm{tot})*}}{\mathrm{i}\hbar} a_{m'}^\dagger a_m. \tag{54b}
$$

Often, a quantum-mechanical formulation is unnecessarily complicated and a classical picture is sufficient. This can be derived in a Hamiltonian framework in exactly the same way as before (Landau and Lifshitz, 1976) if we

1. replace real-space operators a_n, b with classical amplitudes,
2. replace the commutator $[\ldots, \ldots]$ with the Poisson bracket $\{\ldots, \ldots\}$,
3. replace the quantization conditions (40) with an appropriate classical eigenmode normalization.

The first two points are straightforward enough while the third one warrants a short discussion. In quantum mechanics, the modes and operators are normalized such that $a_n^\dagger a_n$ is the photon number inside the quantization volume, i.e., that $\hbar\omega_n L^{-1} a_n^\dagger(z) a_n(z)$ represents the linear energy density in mode n at position z. In contrast, in classical physics one would normalize the eigenmodes that form the basis of the coupled mode theory to some finite unit of energy density $\mathcal{E}_n$ (e.g., 1 J/m) or some unit of power $\mathcal{P}_n$

(e.g., 1 W). The same possibilities also exist in the normalization of the acoustic mode.

From a formal point of view it appears most elegant to apply the same type of normalization to all modes (i.e., normalize optical and acoustic modes to the same unit of energy density or power or at least to the same dimension). However, in practice it is relatively common to use "mixed" normalization schemes where, for example, the optical modes are normalized to a power level whereas the acoustic mode is normalized to a unit of energy density or even displacement amplitude. Although this does not change anything fundamentally in the framework, it is an excellent source of confusion and should be kept in mind when studying an unfamiliar work.

Finally, the classical equations of motion have the advantage that the inclusion of loss is very straightforward. An inspection of Eqs. (54) identifies them as nonlinearly coupled advection equations, to which phenomenological spectral line width constants Γ_n and Γ_b (see, e.g., Wolff et al. (2015b) for a derivation of these constants) can be easily added:

$$(\partial_t + v_n\partial_z + \Gamma_n)a_n = -\mathrm{i} \sum_{m'}^{\omega_n \geq \omega_{m'}} \frac{\omega_n Q_{nm'}^{(\mathrm{tot})}}{\mathcal{E}_n} a_{m'} b - \mathrm{i} \sum_{m}^{\omega_m \geq \omega_n} \frac{\omega_n Q_{mn}^{(\mathrm{tot})*}}{\mathcal{E}_n} a_m b^*, \tag{55a}$$

$$(\partial_t + v_b\partial_z + \Gamma_b)b = -\mathrm{i} \sum_{mm'}^{\omega_m \geq \omega_{m'}} \frac{\Omega_0 Q_{mm'}^{(\mathrm{tot})*}}{\mathcal{E}_b} a_{m'}^* a_m, \tag{55b}$$

when formulated in terms of eigenmodes normalized to energy density. The corresponding equations in power-normalization and using spatial power decay coefficients $\alpha_n = 2\Gamma_n/v_n$ and $\alpha_b = 2\Gamma_b/v_b$ read:

$$\left(\frac{1}{v_n}\partial_t + \partial_z + \frac{\alpha_n}{2}\right)a_n = -\mathrm{i} \sum_{m'}^{\omega_n \geq \omega_{m'}} \frac{\omega_n Q_{nm'}^{(\mathrm{tot})}}{\mathcal{P}_n} a_{m'} b - \mathrm{i} \sum_{m}^{\omega_m \geq \omega_n} \frac{\omega_n Q_{mn}^{(\mathrm{tot})*}}{\mathcal{P}_n} a_m b^*, \tag{56a}$$

$$\left(\frac{1}{v_b}\partial_t + \partial_z + \frac{\alpha_b}{2}\right)b = -\mathrm{i} \sum_{mm'}^{\omega_m \geq \omega_{m'}} \frac{\Omega_0 Q_{mm'}^{(\mathrm{tot})*}}{\mathcal{P}_b} a_{m'}^* a_m. \tag{56b}$$

3.4 Further comments on the derivation

We conclude the derivation of the basic framework with a slightly deeper discussion of mode normalization and the optomechanical interaction terms.

3.4.1 Mode normalization

In the following discussion, we will preempt results from Section 5. In our derivation, we decided to normalize the quantum-mechanical modes to a fixed quantization "volume" L, that could be, e.g., the total length of the waveguide. This allows us to directly connect the quantum theory to the classical theory as illustrated by the identical form, e.g., of Eq. (54a) and Eq. (55a). A minor disadvantage of this choice is that the quantization length L appears explicitly in the equations of motion, when the mode normalization constant $\mathcal{E}_n^{\mathrm{QM}} = \hbar\omega_n/L$ is inserted in Eq. (54a):

$$(\partial_t + v_n\partial_z)a_n = \sum_{m'}^{\omega_n \geq \omega_{m'}} \frac{LQ_{nm'}^{(\mathrm{tot})}}{\mathrm{i}\hbar} a_{m'} + b \sum_{m}^{\omega_m \geq \omega_n} \frac{LQ_{mn}^{(\mathrm{tot})*}}{\mathrm{i}\hbar} a_m b^\dagger, \qquad (57)$$

However, this allows for a nice connection to the theory of Brillouin in resonators, as there the optomechanical interaction is mediated by an optomechanical overlap integral over all three spatial dimensions, divided by Planck's constant (see Eq. 84). In contrast, in a waveguide, the optomechanical overlap is only carried out over the transverse plane (see Eq. 50b) and the additional factor L serves as the missing third integration.

Naturally, it is possible to avoid the introduction of a quantization length within the quantum theory. We discuss this for the optical envelope operators only—the discussion for the acoustic ones is analogous. In essence, it involves dropping the symbol L wherever it appears, i.e., in the definition of the envelope operators (Eq. 38):

$$a_n(z,t) = (2\pi)^{-1/2} \int_{-\infty}^{\infty} \mathrm{d}k \; \exp\left[\mathrm{i}(k - k_n)z - \mathrm{i}\omega_n t\right]\tilde{a}_{nk}(t), \qquad (58)$$

which now have units of $1/\sqrt{\mathrm{m}}$ and satisfy the commutator

$$\left[a_m(z,t), a_n^\dagger(z',t)\right] = \delta_{mn}\delta(z - z'), \qquad (59)$$

if the eigenmodes are normalized to

$$\mathcal{E}_n^{(\mathrm{QM})} = \hbar\omega_n. \qquad (60)$$

Without a unit of length, it is natural to carry z-integrations out over the entire real axis, e.g., in the electromagnetic Hamiltonian:

$$\mathcal{H}^{(\mathrm{EM})} = -\sum_n \int_{-\infty}^{\infty} \mathrm{d}z \; \hbar\left[\mathrm{i}v_n\partial_z + \cdots\right]a_n(z,t)^\dagger a_n(z,t), \qquad (61)$$

which retains units of energy. We obtain the equation of motion without L:

$$(\partial_t + v_n \partial_z) a_n = \sum_{m'}^{\omega_n \geq \omega_{m'}} \frac{Q_{nm'}^{(\mathrm{tot})}}{\mathrm{i}\hbar} a_{m'} + b \sum_{m}^{\omega_m \geq \omega_n} \frac{Q_{mn}^{(\mathrm{tot})*}}{\mathrm{i}\hbar} a_m b^{\dagger}, \tag{62}$$

where the individual coupling constants contributing to $Q_{mn}^{(\mathrm{tot})}$ are still given by the integrals in Eqs. (50a)–(50c). However, as a subtle consequence of this more appealing master equation, all mode-related quantities now have "inconvenient" units. For example the electric mode distribution $\boldsymbol{e}^{(nk_n)}(\boldsymbol{r})$ is now in units of $\mathrm{V}/\sqrt{\mathrm{m}}$—as one easily verifies from Eq. (39a)—and $Q_{mn}^{(\mathrm{tot})}$ has units of $\mathrm{J}\sqrt{\mathrm{m}}$. Ultimately, the exact choice of normalization scheme is a matter of preference.

3.4.2 Equivalence between optical forces and dynamic grating strengths

The derivation of Brillouin scattering from a Hamiltonian as presented above has the particular feature that the optomechanical overlap integrals $Q_{nn'}$ that drive the optical mode equations (54a) naturally show up in the driving terms of the acoustic mode Eq. (54b), as well. In fact, the overlap coefficients $Q_{nn'}$ contain both the deformation-induced modification of the inverse permittivity that couples the different optical modes, as well as the driving of the acoustic amplitude by the optically induced stress field. This can be nicely illustrated by interpreting, e.g., the photoelastic term (Eq. 50b) either as an overlap integral between optical fields and a change in inverse permittivity $\delta\overleftrightarrow{\beta}$:

$$Q_{nn'}^{(\mathrm{PE})} = \int \mathrm{d}^2 r \, \left[d_i^{(nk_n)}\right]^* \underbrace{\frac{p_{ijkl}}{\epsilon_0} s_{kl}^{(q_0)}}_{=\delta\overleftrightarrow{\beta}(\overleftrightarrow{s})} d_j^{(n'k_{n'})},$$

or one between optically induced stress $\overleftrightarrow{t}$ and acoustic strain field:

$$Q_{nn'}^{(\mathrm{PE})} = \int \mathrm{d}^2 r \, \underbrace{\left[d_i^{(nk_n)}\right]^* \frac{p_{ijkl}}{\epsilon_0} d_j^{(n'k_{n'})}}_{=t_{ij}{}^*(\boldsymbol{d}^{(n)}, \boldsymbol{d}^{(n')})} s_{kl}^{(q_0)}.$$

Similar decompositions of the roto-optic and moving-boundary terms are also possible. This equivalence between the two processes is not self-evident from the outset and a consequence of the conservation of energy inherent to a Hamiltonian treatment.

In fact, the equivalence is intimately linked to the Manley–Rowe relations known from nonlinear optics (Boyd, 2008) and more generally to Maxwell relations from thermodynamics. For both types of relations, conservation of energy plays a major role in their applicability. The critical question is whether the interaction is thermodynamically reversible, i.e., whether an optomechanical interaction is the result of a physical process that creates entropy (Wolff et al., 2015b). For example, some optical loss process could contribute an additional force term that is not mirrored on the right-hand side of the optical mode equation. However, in order to drive an SBS-like process with a beat between different optical modes, this can only happen with *nonlinear* optical loss; conventional linear loss due to absorption or scattering will not break the equivalence in this way.

An extreme (albeit not directly SBS-relevant) example of an equivalence-breaking process is the so-called *current striction* observed in some semiconductors such as germanium (Kornreich et al., 1967). Here, the apparent electrostrictive parameter under quasi-dc conditions exceeds the photoelastic parameter by several orders of magnitude. The reason is that in this case the "electrostrictive" stress is in fact not caused by the presence of an electric field itself, but rather the Ohmic current that it drives in the solid. Since this current dissipates energy, the process is irreversible and can contribute stresses that are not mirrored in the photoelastic effect. To the best of our knowledge no such irreversible nonlinear effect has been identified to contribute to a Brillouin process, so far. We should stress that a Hamiltonian framework would become inappropriate in this case.

3.4.3 Optomechanical stress vs force formulation

The electrostrictive term as presented in Eq. (50b) is based on the overlap of the acoustic strain field and an optically induced stress field. The same applies to the roto-optic term in Eq. (50c). The moving-boundary term as presented in Eq. (50a), in contrast, takes the form of the overlap between the acoustic deformation field $\boldsymbol{u}$ and the optical surface pressure, i.e., a force density. This apparently different treatment may seem a little awkward. Some authors might also simply prefer to think in terms of forces rather than stress fields and therefore prefer the former.

The force density is the time derivative of the momentum density and therefore given by the divergence of the momentum flux density, i.e., the stress:

$$\boldsymbol{F} + \mathrm{div}\,\overleftrightarrow{T} = 0. \tag{63}$$

On the other hand, the strain is given by the (symmetrized) gradient of the acoustic displacement field $\boldsymbol{U}$. The photoelastic overlap integral takes the form[j]:

$$Q_{nn'}^{(\mathrm{PE})} = \int \mathrm{d}^2 r\, \overleftrightarrow{t}^{(n,n')} : \left[\mathrm{grad}\, \boldsymbol{u}^{(q_0)}\right], \tag{64}$$

where $\overleftrightarrow{t}^{(n,n')} = \left[d_i^{(nk_n)}\right]^* d_j^{(n'k_{n'})} p_{ijkl}/\epsilon_0$ is the electrostrictive stress. Since $Q_{nn'}^{(\mathrm{PE})}$ is of the form $\int \mathrm{d}^2 r \boldsymbol{f}(\boldsymbol{r}) \cdot \mathrm{grad}\, g(\boldsymbol{r})$, it can be brought to the form $\oint \mathrm{d}s[\hat{n} \cdot \boldsymbol{f}(\boldsymbol{r})]g(\boldsymbol{r}) - \int \mathrm{d}^2 r[\mathrm{div}\, \boldsymbol{f}(\boldsymbol{r})]g(\boldsymbol{r})$ by application of the divergence theorem. This means the photoelastic overlap can also be expressed in the form

$$Q_{nn'}^{(\mathrm{PE})} = \oint \mathrm{d}s\, \left[\hat{n} \cdot \overleftrightarrow{t}^{(n,n')}\right] \cdot \boldsymbol{u}^{(q_0)} + \int \mathrm{d}^2 r\, \underbrace{\left[-\mathrm{div}\, \overleftrightarrow{t}^{(n,n')}\right]}_{=\boldsymbol{f}^{(n,n')}} \cdot \boldsymbol{u}^{(q_0)}, \tag{65}$$

where $\boldsymbol{f}^{(n,\,n')}(\boldsymbol{r})$ is the electrostrictive force density, the contour integral is carried out over the boundary of each electrostrictive material and $\hat{n}$ is the local normal vector of the integration contour. This means it is possible to formulate the electrostrictive interaction in terms of forces rather than stress fields, but at the expense of an additional boundary term. In the end, it is very much a matter of taste which formulation one finds more elegant or useful—within this chapter we stick with the stress–strain formulation.

3.4.4 Roto-optic and artificial interaction terms

Originally, Brillouin physics was almost entirely studied in homogeneous, optically mostly isotropic media. As a result, for a very long time the only relevant optomechanical interaction process was photoelasticity and electrostriction. This means that, for most of its history, Brillouin scattering was dominated by the bulk stresses caused by the optical intensity.

With the miniaturization of photonics and especially the wide availability of silicon nanowires as photonic waveguides, surface pressures have gained importance as an additional source of optomechanical interaction via the moving boundary and radiation pressure term, which can enhance

[j] The keen reader will spot the absence of the symmetrized gradient. This is not a problem, because the stress tensor is symmetric and therefore the antisymmetric part of the gradient does not contribute to the total expression.

or inhibit the photoelastic coupling and even cancel it for certain acoustic modes (Florez et al., 2016).

In a similar vein, we note that optical metamaterials provide a large range of artificial media with strong optical and mechanical anisotropy. This opens the door to add a third optomechanical interaction process via the bulk *torque* density induced by the optical fields described by the roto-optic tensor r_{ijkl}. Unlike the photoelastic tensor p_{ijkl}, which is a phenomenological parameter of a material unrelated to other material properties, the roto-optic tensor is entirely defined by the dielectric tensor components (Nelson and Lax, 1970) (see Eq. 31). This opens a third possibility to tailor (enhance or cancel) the interaction between optical and acoustic modes via the inclusion of anisotropic metamaterials composed of materials with strong dielectric and mechanical contrasts, whose effective photoelastic and roto-optic tensors can exceed those of naturally occurring materials by orders of magnitude (Smith et al., 2017).

3.5 Thermal acoustic noise

Noise in Brillouin scattering arises predominantly from the "bath" of thermal phonons that exist in all noncryogenic experiments. Indeed, taking a nominal acoustic phonon frequency of $\Omega/(2\pi) = 7$ GHz, we can expect a room temperature thermal occupation of around $k_B T/(\hbar\Omega) \approx 900$ phonons per mode. This noise initiates the Brillouin process in the "spontaneous" (pump only) configuration, and incoherent scattering from thermal phonons can be amplified by existing Brillouin gain processes in a manner similar to amplified spontaneous emission in a semiconductor optical amplifier. Because most Brillouin experiments operate in the strongly coherent regime, the modeling of noise processes is not often required, however, noise can become important when analyzing the physical limits of Brillouin optical information processing and acoustic storage.

To model the noise, one can in principle work from the quantum formalism in the style of Drummond and Corney (2001), writing down a Hamiltonian for the coherent fields coupled weakly to a thermal reservoir; it is possible then to derive quantum Langevin equations for the optical and acoustic envelopes. An alternative approach (Nieves et al., 2021a) is to include the noise empirically as a driving term to the acoustic envelope equation (54b), which then becomes

$$(\partial_t + v_b \partial_z + \Gamma_b) b = -\mathrm{i} \sum_{mm'}^{\omega_m > \omega_{m'}} \frac{\Omega_0 Q_{mm'}^{(\mathrm{tot})*}}{\mathcal{E}_b} a_{m'}^* a_m + \sqrt{\frac{\sigma}{\mathcal{E}_b}} R(z, t), \tag{66}$$

where $R(z, t)$ is a complex-valued white-noise term, with zero mean and delta-correlations in space and time. The strength of the noise σ is connected to the elastic viscosity via the fluctuation-dissipation theorem:

$$\sigma = 2k_B T\Gamma_b, \tag{67}$$

with k_B being Boltzmann's constant and T being the temperature. Eq. (66) can be solved numerically (or, under some circumstances, analytically) using standard methods for stochastic partial differential equations (see, e.g., Nieves et al., 2021b).

4. Typical settings

4.1 Backward SBS

The CMEs so far have been derived for an arbitrary number of optical modes without a priori information about the relative frequencies and directions of propagation. In practice, one will of course specialize them to the problem at hand. To illustrate this, we now consider a Backward SBS (BSBS) example with two optical modes—a pump a_P and a Stokes a_S with $\omega_P > \omega_S \approx \omega$ and $v_P = v_0 = -v_S$. Inserted into the above results, we find the familiar BSBS equations in quantum form:

$$(\partial_t + v_0\partial_z)a_P(z,t) = -\mathrm{i}\frac{\omega_P Q_{PS}^{(\mathrm{tot})}}{\mathcal{E}_P^{(\mathrm{QM})}} a_S(z,t)b(z,t), \tag{68a}$$

$$(\partial_t - v_0\partial_z)a_S(z,t) = -\mathrm{i}\frac{\omega_S Q_{PS}^{(\mathrm{tot})*}}{\mathcal{E}_S^{(\mathrm{QM})}} a_P(z,t)b^\dagger(z,t), \tag{68b}$$

$$(\partial_t + v_b\partial_z)b(z,t) = -\mathrm{i}\frac{\Omega Q_{PS}^{(\mathrm{tot})*}}{\mathcal{E}_b^{(\mathrm{QM})}} a_S^\dagger(z,t)a_P(z,t), \tag{68c}$$

and in the classical power formulation, where we assume $\alpha_1 = -\alpha_2 = \alpha_0$ and normalize both optical modes to the same power level $\mathcal{P}_0$:

$$\left(\frac{1}{v_0}\partial_t + \partial_z + \frac{\alpha_0}{2}\right)a_P(z,t) = -\mathrm{i}\frac{\omega Q_{PS}^{(\mathrm{tot})}}{\mathcal{P}_0} a_S(z,t)b(z,t), \tag{69a}$$

$$\left(\frac{-1}{v_0}\partial_t + \partial_z - \frac{\alpha_0}{2}\right)a_S(z,t) = \mathrm{i}\frac{\omega Q_{PS}^{(\mathrm{tot})*}}{\mathcal{P}_0} a_P(z,t)b^*(z,t), \tag{69b}$$

$$\left(\frac{1}{v_b}\partial_t + \partial_z + \frac{\alpha_b}{2}\right)b(z,t) = -\mathrm{i}\frac{\Omega Q_{PS}^{(\mathrm{tot})*}}{\mathcal{P}_b} a_S^*(z,t)a_P(z,t). \tag{69c}$$

We skip the classical energy formulation for brevity.

The BSBS equations of motion have no closed analytical solution and in general cases such as experiments with short interacting pulses (Piotrowski et al., 2021) they must be solved numerically. However, it is well known that within several quite restrictive approximations an analytical solution can be found. This is firstly the assumption of a steady state, which lets us drop all time derivatives. Furthermore, it involves the local acoustic response approximation (also known as a Markovian approximation for the acoustic field), which allows us to eliminate the acoustic field (see, e.g., Wolff et al., 2015b) and absorb it into a coefficient

$$G = \frac{4\omega\Omega|Q_{PS}|^2}{\mathcal{P}_0^2\mathcal{P}_b\alpha_b}, \tag{70}$$

resulting in equations for the optical power levels $P_P = \mathcal{P}_P|a_P|^2$ and $P_S = \mathcal{P}_S|a_S|^2$ for pump and Stokes, respectively:

$$\partial_z P_p = -\alpha_0 P_p - G P_p P_s, \qquad \partial_z P_s = \alpha_0 P_s - G P_p P_s, \tag{71}$$

which can be easily solved in the absence of optical loss ($\alpha_0 = 0$) by reformulating in terms of sum and difference powers $P_+ = P_p + P_s$, $P_- = P_p - P_s$:

$$\partial_z P_+ = -\frac{G}{2}(P_+^2 - P_-^2), \qquad \partial_z P_- = 0. \tag{72}$$

The second equation states that P_- is a constant while the first equation can be solved, being of the Riccati type. A typical solution is shown in Fig. 2. It perfectly illustrates the characteristic feature of BSBS: an energy transfer from the pump to the counter-propagating Stokes.

4.2 Forward Brillouin

The second very common setting for Brillouin scattering experiments is the excitation of a higher order acoustic wave with very small forward wave number $q_0 \approx 0$ (very often referred to as a transverse acoustic mode) by the interference between different Stokes orders of the same optical mode. For the optical side bands, we use the Stokes order as the index, i.e., $n = 0$ is the central pump wave, $n = 1$ and $n = -1$ are the first Stokes and anti-Stokes waves, respectively. We assume that all parameters such as optical carrier, group velocity and coupling parameter are the same: $\omega_1 \approx \omega_2 \approx \cdots = \omega$, $v_1 = v_2 = \cdots$ and $Q_{n(n-1)}^{(\mathrm{tot})} = Q^{(\mathrm{tot})}$. As before, we assume steady state,

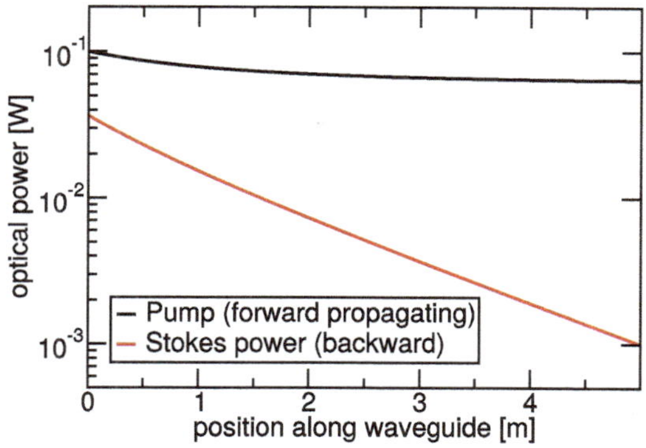

Fig. 2 Typical behavior of narrow-band (quasi-stationary) BSBS: A weak Stokes (*red line*) propagating to the left gains energy from back-scattered pump light (*black line*). The system is a 5 m long waveguide with a power gain of $G = 10\mathrm{W}^{-1}\mathrm{m}^{-1}$; the injected powers were 100 mW and 1 mW for pump and Stokes, respectively. The total Stokes gain experienced is approx. 16 dB, the pump depletion 2 dB.

absence of optical loss and a local acoustic response, which is even more justified than in BSBS, because FBS usually involves a near stationary acoustic mode. Finally, we assume that the acoustic mode always couples "adjacent" Stokes orders, i.e., $n = 0$ to $n = \pm 1$. The governing equations then read:

$$\partial_z a_n = -\frac{\mathrm{i}\omega}{\mathcal{P}_0}\left(Q^{(\mathrm{tot})} a_{n+1} b + Q^{(\mathrm{tot})^*} a_{n-1} b^*\right), \tag{73a}$$

$$b = -\frac{\mathrm{i}\Omega Q^{(\mathrm{tot})*}}{\mathcal{E}_b \Gamma_b} \sum_n a_n^* a_{n+1}, \tag{73b}$$

where the optical group velocity v disappeared from the equations when the time derivatives was neglected in the assumption of a steady state (compare with Section 4.1).

These equations are formulated over infinitely many (anti-)Stokes orders $-\infty < n < \infty$ and one would be tempted to restrict the values of n taken in account as much as practically possible.

However, there is the danger to overdo this simplification and—inspired by the BSBS case—just assume a pump and a Stokes, which following the same steps and further approximations as above would lead to:

$$\partial_z P_p = -G P_s P_p, \quad \text{note: inappropriate model!} \tag{74a}$$

$$\partial_z P_s = G P_p P_s, \quad \text{note: inappropriate model!} \tag{74b}$$

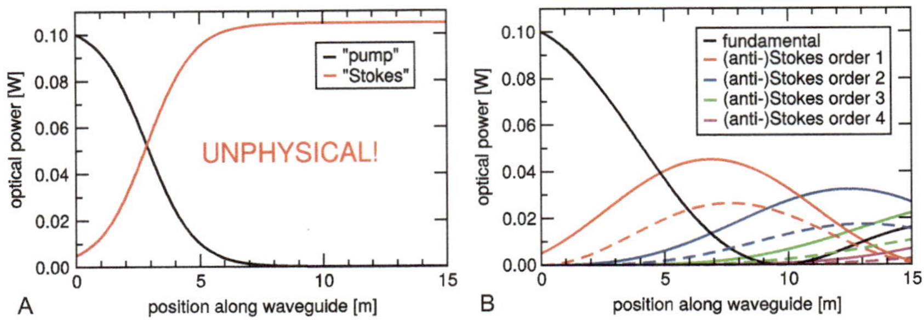

Fig. 3 Contradictory predictions of the behavior of narrow-band (quasi-static) forward Brillouin scattering from differently truncated sets of coupled mode equations: (A) incorrectly predicted unidirectional power transfer from pump to Stokes due to incorrect truncation of Stokes orders in a BSBS-like fashion; (B) physically correct symmetric redistribution of energy between Stokes and anti-Stokes modes as well as the onset of cascading. Stokes and anti-Stokes powers of the same order are shown in the same color as *solid* and *dashed lines*, respectively. Note in particular how the (fundamental) "pump" picks up energy again from $z = 10$m onwards. The "gain" parameter is $G = 10\text{W}^{-1}\text{m}^{-1}$, waveguide length 15m and injected power are 100 and 5mW for the $n = 0$ ("pump") and $n = 1$ ("Stokes") fields, respectively.

where the parameter

$$G = \frac{2\omega\Omega |Q^{(\text{tot})}|^2}{\mathcal{P}_0^2 \mathcal{E}_b \Gamma_b} \tag{74c}$$

is formally equivalent to the BSBS gain in Eq. (70). These equations describe a unidirectional transfer of energy from the pump to the Stokes as shown in Fig. 3A. This result is completely wrong and a pure artifact of inappropriately truncating the Stokes orders.

The problem is that not only one pump and corresponding Stokes are phase-matched to a given acoustic mode as is the case in BSBS. Instead, the Stokes itself can again act as a pump for the second-order Stokes and the original pump as the Stokes of the first-order anti-Stokes. This means that FBS is inherently cascaded and the correctness of a simulation should be at the very least checked by increasing the set of (anti-)Stokes orders in both directions. Fig. 3B shows the results for Eqs. (73) that using (anti-)Stokes orders $-10 \leq n \leq 10$ (only up to $\pm$ 4 shown).

The result is fundamentally different from the oversimplified model and ultimately highlights that FBS in its basic form leads to a phase modulation instead of an amplitude modulation as in the case of BSBS (Wolff et al., 2017). This is also the reason that while Eqs. (70) and (74c) both describe the energy transfer between optical bands and have the same form, only the former actually constitutes a *gain* parameter.

4.3 Intermodal Brillouin

We have discussed the two most common settings of Brillouin—namely intramodal BSBS and nondispersive intramodal FBS. One key insight is that they behave extremely differently: strictly two interacting optical bands and pure amplitude modulation in the BSBS case; strictly symmetric power distribution and pure phase modulation in the FBS case. This naturally leads to the question: are there intermediate regimes connecting these two obvious extremes? We will answer this in the following by analyzing different intermodal cases and briefly study the role of optical mode dispersion.

4.3.1 Symmetry aspects, problem classification

When moving from the two already discussed settings (BSBS and FBS), which also both happen to be intramode processes, to the more general intermode settings, there are mostly two aspects to worry about. First and unlike in the intraband case, the two optical modes will generally have different symmetry. This means that different "selection rules" apply as to which acoustic modes have appreciable optomechanical coupling. In all but the most asymmetric waveguide geometries, this can be analyzed with group theory; in the case of round waveguides this boils down to conservation of total angular momentum. A collection of some common symmetry classes can be found in Wolff et al. (2014). As a general rule of thumb, intramodal Brillouin processes will usually involve acoustic modes with longitudinal or predominantly radial deformation fields. In contrast, in the case of intermodal Brillouin processes, the acoustic field must "compensate" the symmetry mismatch of the two participating optical modes and will often have strong shear or rotational components. For example, the two fundamental optical modes of a birefringent waveguide will be coupled to each other via a rocking mode.

Second, the overall way the Brillouin process behaves (e.g., Stokes amplification vs pure phase modulation) is determined by the relative positions of the participating optical modes within the phase-matching diagram (see Fig. 1.5 in chapter "Historical perspective and basic principles" by Poulton et al.) This means that processes with a similar *relative* arrangement of optical wave numbers and energies will exhibit similar dynamics.

As it turns out, we can classify all forward or backward, intramode or intermode Brillouin settings broadly into three categories, that will be discussed in the following.

4.3.2 Nondispersive degenerate forward cases

The first class is that where both the Stokes and anti-Stokes process involve the same acoustic mode with the same wavenumber, connecting optical carriers on the same optical dispersion line. This process can be both intramodal, i.e., pump and (anti-)Stokes are the same optical mode or intermodal, i.e., pump and (anti-)Stokes are optical modes that are on degenerate bands, but have orthogonal mode profiles. The former case means the same optical mode with the same polarization, the latter case implies that all optical modes have the same relationship between frequency and wavenumber, but different symmetry—e.g., orthogonal polarizations.

In all these cases, we can expect the system to behave similarly to the basic FBS case outlined in Section 4.2, because the same acoustic mode is equally strongly involved in distributing energy from the pump to the Stokes and the anti-Stokes; it is equally strongly amplified in the former and depleted in the latter process. The result is a tendency for cascaded scattering and phase modulation of the input signal. For the modeling, this implies that great care must be taken not to truncate the (anti-)Stokes orders too aggressively lest one risks unphysical theoretical predictions.

4.3.3 Strongly nondegenerate forward and backward intermodal cases

The second main class is where the Stokes and anti-Stokes processes involve acoustic waves with clearly distinct wave numbers, usually $q_S \approx -q_{AS}$, where q_S and q_{AS} belong to the Stokes-active and the anti-Stokes-active acoustic modes, respectively. The most prominent example for this class is clearly classic backward SBS as discussed in Section 4.1, where $q_S \approx 2k_P \approx -q_{AS}$. Further examples include counter-propagating optical modes from different branches of the dispersion relation (e.g., different polarization) as well as *forward* propagating optical modes whose mode indices are sufficiently different, e.g., in strongly birefringent waveguides or between fundamental and higher order optical modes (see Fig. 4).

In these settings, the Stokes and anti-Stokes process interact with completely separate acoustic excitations, which means that the one participating in the Stokes process is continually amplified while the other one is continually depleted by the upconversion of optical excitations. This means that this class of problems will generally behave similarly to conventional

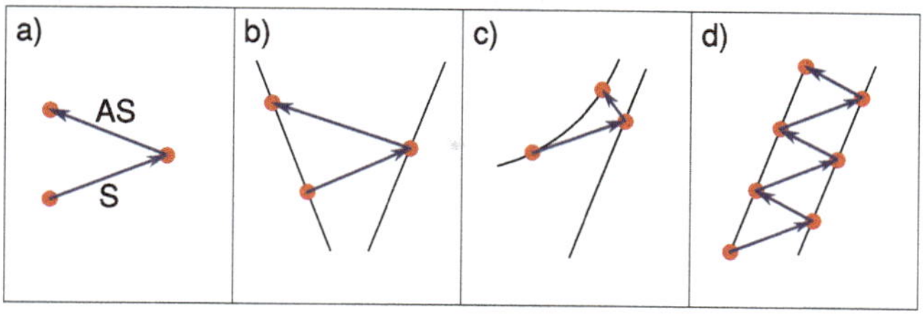

Fig. 4 Phase-matching diagrams for different Brillouin processes that share a general similarity with backward SBS. *Red dots* indicate carrier wave numbers and frequencies of the participating optical modes on their respective dispersion relations (shown in *black*). *Blue arrows* indicate wave numbers and frequencies of the acoustic modes phase-matched to them—the labels 'S' and 'AS' denote the phonons enabling the Stokes and anti-Stokes processes, respectively. (A) The general case: Stokes and anti-Stokes phonons are strongly distinct, typically with wave numbers of opposite sign. (B) Most common realization: backward SBS, both intramodal and intermodal. (C) Intermodal *forward* scattering that behaves rather like BSBS, because it occurs between two strongly nondegenerate optical modes. (D) Corner case: intermodal forward scattering involving two parallel branches. While Stokes and anti-Stokes process use clearly different acoustic modes, scattering from pump to second Stokes (via the first Stokes) involves the same acoustic modes as scattering to the second anti-Stokes, so cascading and phase modulation is permitted by phase matching. Such cases must be modeled very carefully to avoid incorrect theoretical predictions.

backward SBS with dominant transfer of power from the pump to the (first) Stokes. A special case is that of forward scattering between two parallel optical branches, where cascading between even order (anti-)Stokes lines is permitted by the phase-matching condition and which should be modeled carefully with this possibility in mind. Otherwise, a rather simple model involving just the pump and the Stokes should suffice in many cases.

4.3.4 Dispersive degenerate and weakly birefringent forward cases

So far, we have identified more general settings that still behave mostly like the extreme cases in Sections. 4.3.2 and 4.3.3. However, one would expect a regime where the character gradually shifts from BSBS-like to FBS-like. From the previous discussion it should be clear that this cross-over regime appears when the Stokes and anti-Stokes process involve acoustic waves with *almost*, but not quite the same wave number $q_S \approx q_{AS}$. There are two notable examples for this situation: degenerate (inter- or intramodal) FBS with optical mode dispersion; and intermodal FBS between almost degenerate branches. As before, the relative position of the optical modes in the phase-matching diagram is more important for the overall qualitative

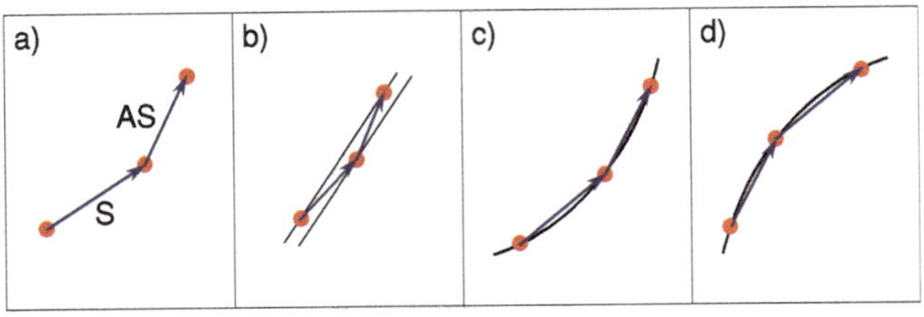

Fig. 5 Phase-matching diagrams similar to Fig. 4 for different Brillouin processes in the intermediate regime between BSBS-like and FBS-like dynamics. (A) The general case: Stokes and anti-Stokes phonon are not quite the same, but close enough that they can participate off-resonantly in the respective opposite effect. (B) intermodal forward scattering between almost degenerate optical modes. This is the obvious cross-over regime between the extreme cases discussed in Sections 4.3.2 and 4.3.3. (C) Mode dispersion in both intramodal and degenerate intermodal FBS can play a similar role to weak birefringence. (D) Note how positive and negative mode dispersion correspond to selecting the pump on different branches in case (B).

behavior than precisely which type of dispersion relation led to this arrangement (Fig. 5).

The impact of weak dispersion on cascaded FBS was studied in the past (Wolff et al., 2017) and can be summarized as follows: If cascaded FBS leads to a pure phase modulation without change of the intensity variations along the waveguide, then dispersion gradually shifts the relative phases between the different Stokes orders and will partially transform the phase modulation into an amplitude modulation. Depending on the sign of the dispersion parameter, this can either act to increase or decrease the contrast of optical intensity fluctuations along the waveguide. First, this translates to an increased (or decreased) dynamic grating contrast and resulting increased (decreased) modulation efficiency along the waveguide. Second, the fact that the acoustic grating is sustained by interfering optical waves implies that the spectrum experiences an overall redshift along the waveguide. This redshift will become more pronounced if the grating becomes stronger along the waveguide.

5. Brillouin in optical resonators

So far we explored optical and elastic waveguide geometries with continuous translational symmetry along the propagation direction. Within such waveguides, the Brillouin interaction was described through the spatio-temporal evolution of slowly varying optical, $a_n(z, t)$, and mechanical, $b(z, t)$, field amplitudes, as given by Eqs. (54). In this section, our goal is to

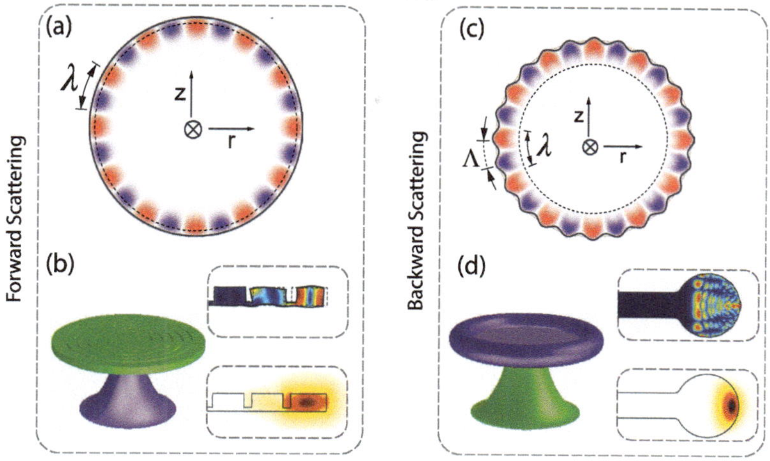

Fig. 6 Illustration of mechanical modes phase-matched for forward (A, B) and backward (C, D) Brillouin scattering in axisymmetric cavities. The colormaps insets in (B, D) represent the electric field, *blue* (*red*) colors represent negative (positive) values; λ (Λ) is the optical (mechanical) wavelength. While in (A) the cavity expands radially with no azimuthal distortion, in (B), the curved boundary represents the short-wavelength mechanical wave. *Adapted from Wiederhecker, G.S., Dainese, P., Mayer Alegre, T.P., 2019. Brillouin optomechanics in nanophotonic structures. APL Photonics 4 (7), 071101. https://doi.org/10.1063/1.5088169.*

derive similar equations, but describing the time evolution of discrete optical and mechanical resonator modes coupled through the Brillouin interaction. In Fig. 6, we illustrate two resonator geometries in which forward and backward Brillouin scattering have been experimentally observed (Honda et al., 2018; Santos et al., 2017). The coupled wave model derived in this section describes the Brillouin interaction observed in these and other resonator devices (Grudinin et al., 2009, 2010; Otterstrom et al., 2018). A more thorough discussion on the specific settings of the Brillouin interaction are given in chapter "The convergence of cavity optomechanics and Brillouin scattering" by Schmidt et al.

5.1 Cavity optomechanical Hamiltonian

To obtain a Hamiltonian suitable for optical cavity modes, we employ cavity field operators in the eigenmode expansion of Eqs. (36),

$$\boldsymbol{D}(\boldsymbol{r}, t) = \sum_n a_n(t)\boldsymbol{d}^{(n)}(\boldsymbol{r}) + \text{h.c.}, \tag{75a}$$

$$\boldsymbol{U}(\boldsymbol{r}, t) = b(t)\boldsymbol{u}(\boldsymbol{r}) + \text{h.c.}, \tag{75b}$$

where the operators are normalized such that the $a_n^\dagger a_n$ and $b^\dagger b$ are dimensionless and their expectation values represent average photon and phonon numbers. Because the elastic Hamiltonian involves the strain field, it is useful to employ a strain modal expansion based on Eq. (75b),

$$S_{ij}(\boldsymbol{r},t) = b(t)s_{ij}(\boldsymbol{r}) + \text{h.c.}, \tag{76}$$

where the modal strain spatial profile is obtained according to the strain definition in Eq. (4). These operators should follow (equal-time) commutation relations analogous to Eqs. (40),

$$[a_m(t), a_n^\dagger(t)] = \delta_{mn}, \tag{77a}$$

$$[b(t), b^\dagger(t)] = 1. \tag{77b}$$

Substituting the modal expansion for electric displacement, Eq. (75a), and the analogous magnetic flux density $\boldsymbol{B}(\boldsymbol{r}, t)$ expansion in the respective Hamiltonians (Eqs. 22 and 28), we derive the following quantized Hamiltonians (up to a constant energy term) (Bhat and Sipe, 2006; Sipe and Steel, 2016; Sipe et al., 2004),

$$\mathcal{H}^{(\text{EM})} = \sum_n \hbar\omega_n\, a_n^\dagger(t)a_n(t), \tag{78a}$$

$$\mathcal{H}^{(\text{mech})} = \hbar\Omega\, b^\dagger(t)b(t). \tag{78b}$$

Note that we have not included energy coefficients (e.g., $\sqrt{\hbar/\omega_n/2}$) in the expansions of Eqs. (75). Therefore, the coefficients $\hbar\omega_n$ and $\hbar\Omega$ in Eqs. (78) are due to our mode normalization choice,

$$\frac{2}{\varepsilon_0}\int \mathrm{d}^3 r d_i^{(n)}(\boldsymbol{r})^* \beta_i j(\boldsymbol{r}) d^{(n)}(\boldsymbol{r})_j^* = \hbar\omega_n, \tag{79a}$$

$$2\Omega^2 \int \mathrm{d}^3 r \rho(\boldsymbol{r})|\boldsymbol{u}^{(n)}(\boldsymbol{r})|^2 = \hbar\Omega. \tag{79b}$$

From these Hamiltonians and the commutation relations, we can easily obtain the Heisenberg time evolution equations for each operator (see Eq. 26),

$$\frac{\partial}{\partial t}a_n = -\mathrm{i}\omega_n a_n \rightarrow a_n(t) = a_n(0)\exp(-\mathrm{i}\omega_n t), \tag{80a}$$

$$\frac{\partial}{\partial t}b = -\mathrm{i}\Omega b \rightarrow b(t) = b(0)\exp(-\mathrm{i}\Omega t). \tag{80b}$$

These solutions for the noninteracting operators motivate the arguments we shall use to ignore specific contributions to the interaction Hamiltonian in the forthcoming discussion. We focus only on the photoelastic and moving boundary contributions, $\mathcal{H}^{(\mathrm{OM})} = \mathcal{H}^{(\mathrm{PE})} + \mathcal{H}^{(\mathrm{MB})}$ as they are the most relevant in experiments.[k] For instance, $\mathcal{H}^{(\mathrm{PE})}$ is obtained by substituting the modal expansions Eqs. (75a) and (76) in Eq. (35a). Similarly to the waveguide counterpart, Eq. (45), the photoelastic energy density is given by

$$\mathcal{I}^{(\mathrm{PE})} = \frac{1}{2\varepsilon_0}\sum_{n,m}\left[a_n d_i^{(n)} + \mathrm{h.c.}\right]\left[a_m d_j^{(m)} + \mathrm{h.c.}\right]\left[b\, p_{ijkl}\, s_{kl} + \mathrm{h.c.}\right], \tag{81}$$

where we have dropped the temporal dependence of the field operators $(a_n(t), b(t))$ and the spatial dependence of the profiles $(d_i(\boldsymbol{r}),\, s_{ij}(\boldsymbol{r}),\, p_{ijkl}(\boldsymbol{r}))$ to simplify the notation. Carrying out the multiplications in the summand of Eq. (81) leads to eight terms. The relevance of each depends on whether their time-dependence is resonant with the time evolution of the noninteracting operators, given by Eq. (80). For instance, terms involving products such as $a_n a_m$ or $a_n^\dagger a_m^\dagger$ will oscillate at $\pm(\omega_n + \omega_m)$ and will be negligible in the dynamics (off-resonance). Note that we ignore the mechanical frequency contribution because $\Omega \ll \omega_{n,\,m}$. This argument leaves only four terms in the summand of Eq. (81), which are proportional to the operator products: $a_n a_m^\dagger (b + b^\dagger)$ and $a_n^\dagger a_m (b + b^\dagger)$. In order to fully explore the commutation relations, we separate the sum in Eq. (81) into intramode $(n = m)$ and intermodal terms $(n \neq m)$:

$$\begin{aligned}
\mathcal{I}^{(\mathrm{PE})} = {} & \frac{p_{ijkl}}{\varepsilon_0}\sum_{n}\left\{a_n^\dagger a_n b\left[d_i^{(n)} s_{kl} d_j^{(n)*}\right] + a_n^\dagger a_n b^\dagger\left[d_i^{(n)} s_{kl}^* d_j^{(n)*}\right]\right\} \\
& + \frac{p_{ijkl}}{2\varepsilon_0}\sum_{n,\,m}^{m\neq n}\left\{a_n a_m^\dagger b\left[d_i^{(n)} s_{kl} d_j^{(m)*}\right] + a_n a_m^\dagger b^\dagger\left[d_i^{(n)} s_{kl}^* d_j^{(m)*}\right]\right\} \\
& + \frac{p_{ijkl}}{2\varepsilon_0}\sum_{n,\,m}^{m\neq n}\left\{a_n^\dagger a_m b\left[d_i^{(n)*} s_{kl} d_j^{(m)}\right] + a_n^\dagger a_m b^\dagger\left[d_i^{(n)*} s_{kl}^* d_j^{(m)}\right]\right\},
\end{aligned} \tag{82}$$

where we used $a_n a_n^\dagger = 1 + a_n^\dagger a_n$ for the intramode summand and ignored the constant term that does not contribute to the dynamics. Similarly to the discussion for waveguides in Section 3.2, in the specific case of the

[k] Note that recent experiments demonstrate backward SBS with strong contribution from exotic effects, such as fountain pressure in superfluid helium (He et al., 2020), which are not covered by the interaction Hamiltonians presented in this chapter.

moving boundary interaction, such constant term can be linked to Casimir-like vacuum forces. To further manipulate the $(m \neq n)$ sums, one should keep in mind that Eq. (82) is also being summed over the repeated indexes (i, j). All operators commute in these sums and the Pockels tensor is symmetric $(p_{ijkl} = p_{jikl})$(Yariv et al., 1984), therefore, the terms proportional to either b or $b^\dagger$ may be combined and the factors 1/2 in Eq. (82) eliminated. Finally, the photoelastic Hamiltonian is obtained by integration of Eq. (82),

$$\mathcal{H}^{(\mathrm{PE})}/\hbar = \sum_{n}\left[a_n^\dagger a_n\left(b g_{n,n}^{(\mathrm{PE})} + \mathrm{h.c.}\right)\right] + \sum_{n,\,m}^{m \neq n}\left[a_n^\dagger a_m b g_{n,m}^{(\mathrm{PE})} + \mathrm{h.c.}\right], \tag{83}$$

where the coupling coefficients are given by

$$g_{n,m}^{(\mathrm{PE})} = \frac{1}{\hbar\varepsilon_0}\int \mathrm{d}^3 r\left({d_i^{(n)}}^{*} p_{ijkl} s_{kl} d_j^{(m)}\right), \tag{84}$$

and have SI units of [Hz]. In contrast to Eq. (47), we do not restrict the sum to $\omega_n > \omega_m$, this is so because there are some experimental situations in Brillouin cavity optomechanics where the $\omega_n < \omega_m$ terms are important due to the low frequency of the elastic waves (He et al., 2020).

The moving boundary interaction Hamiltonian $\mathcal{H}^{(\mathrm{MB})}$ (Eq. 35c) can be derived following the same procedure, leading to an expression identical to Eq. (83), but with the coupling coefficient [units of (Hz)] given by the integral over all cavity dielectric interfaces

$$\begin{aligned} g_{n,m}^{(\mathrm{MB})} = \frac{1}{\hbar}\int_S \mathrm{d}^3 r\, [\hat{n}\cdot \boldsymbol{u}] \times \Big\{ &(\epsilon_a - \epsilon_b)\epsilon_0 [\hat{n}\times \boldsymbol{e}^{(n)}]^{*} \cdot [\hat{n}\times \boldsymbol{e}^{(m)}] \\ &-(\epsilon_a^{-1} - \epsilon_b^{-1})\epsilon_0^{-1}[\hat{n}\cdot \boldsymbol{d}^{(n)}]^{*}[\hat{n}\cdot \boldsymbol{d}^{(m)}]\Big\}, \end{aligned} \tag{85}$$

where ϵ_a (ϵ_b) is the lower (higher) refractive index material and the surface normal $\hat{n}$ points from ϵ_b to ϵ_a. Consistently picking such surface normal orientation is critical when the optical mode interacts with multiple dielectric interfaces. It is also worth mentioning that in the case of backward Brillouin scattering, e.g., in whispering gallery resonators, the backward propagating optical fields are not identical to the forward ones, which often leads to a smaller coupling rate (Espinel et al., 2017).

Therefore, the total cavity optomechanical Hamiltonian—disregarding the roto-optic contribution (Eq. 35b)—is given by

$$\mathcal{H}/\hbar = \Omega b^\dagger b + \sum_{n}\omega_n a_n^\dagger a_n + \underbrace{\sum_{n,\,m}\left[a_n^\dagger a_m b g_{n,m} + \mathrm{h.c.}\right]}_{\mathcal{H}^{(\mathrm{OM})}}, \tag{86}$$

where $g_{n,m} = g_{n,m}^{(\mathrm{PE})} + g_{n,m}^{(\mathrm{MB})}$ and the sum is performed over all indexes, including thus self and cross interaction terms. Note that these coupling coefficients may have opposite signs and care should be taken to avoid incorrect prediction of the net coupling strength (Florez et al., 2016). Wiederhecker et al. (2019) provide a detailed discussion and examples on the evaluation of the coupling coefficients for both waveguide and cavity geometries.

It is also worth mentioning that the $g_{n,m}^{(\mathrm{MB,PE})}$ are in general complex numbers, but it is possible to perform a global phase rotation of the eigenmodes to transform them to real-valued numbers. This is the case when optical and elastic materials are lossless and the inner product defined in Eqs. (14) and (20) applies. A situation where such an inner product is not readily applicable is when dissipative optomechanical interactions occur (Primo et al., 2020). Nevertheless, when using modal fields obtained through numerical methods, such as finite elements, it is often easier to employ the complex-valued expressions derived above.

5.2 Equations of motion

The Heisenberg equations (Eq. 26) may now be obtained using the appropriate commutation relations. For instance, the cross-term interaction in Eq. (86) is calculated as

$$\begin{aligned}
[\mathcal{H}^{\mathrm{OM}}, a_n] &= \hbar \sum_{l,m}^{l\neq m} \left([a_l^\dagger a_m b g_{l,m}, a_n] + [a_l a_m^\dagger b^\dagger g_{lm}^*, a_n] \right) \\
&= \hbar \sum_{l,m}^{l\neq m} \left(b g_{lm} \underbrace{[a_l^\dagger a_m, a_n]}_{=-\delta_{l,n}} + b^\dagger g_{l,m}^* \underbrace{[a_l a_m^\dagger, a_n]}_{=-\delta_{mn}} \right) \\
&= -\hbar \sum_{m}^{m\neq n} \left(g_{n,m} b + g_{m,n}^* b^\dagger \right) a_m.
\end{aligned} \tag{87}$$

Calculating the remaining commutators, the equations of motion are obtained

$$\frac{\partial}{\partial t} a_n = -\mathrm{i}\omega_n a_n - \mathrm{i} \sum_m \left(g_{n,m} b + g_{m,n}^* b^\dagger \right) a_m, \tag{88a}$$

$$\frac{\partial}{\partial t} b = -\mathrm{i}\Omega b - \mathrm{i} \sum_{n,m} \left(g_{n,m}^* a_m^\dagger a_n \right), \tag{88b}$$

where the sums now include all combinations of (m, n). In this form, these equations can be readily applied in the description of intramode Brillouin scattering $(m = n)$, which is the typical scenario in the field of single-mode cavity optomechanics, but also can be extended to the more general intermodal case, where the $(m \neq n)$ terms are relevant.

5.2.1 Optical drive and dissipation

One crucial aspect that is not captured by Eq. (88) is the finite photon and phonon lifetimes. Such loss channels lead to irreversible processes that requires the theoretical framework of open quantum systems. A thorough discussion of such methods, with emphasis on the optomechanical interaction, can be found in (Agarwal, 2013; Bowen and Milburn, 2015; Clerk et al., 2010). Here we limit ourselves to derive the addition of a coherent pump by adding the following Hamiltonian,

$$\mathcal{H}^{(\mathrm{drive})} = i\hbar(E_\mathrm{p} \exp(-i\omega_\mathrm{d} t) a_\mathrm{p}^\dagger + \mathrm{h.c.}), \tag{89}$$

where $E_\mathrm{p} = \sqrt{\kappa_e \mathcal{P}/\hbar\omega_\mathrm{d}}$ is classical driving strength written in terms of the optical pump power $\mathcal{P}$ (units of [W]), ω_d is the pump drive frequency (close to the target cavity mode frequency ω_p), κ_e is the external coupling rate between the bus waveguide and the resonator mode. Note that κ_e depends on the modal overlap between the specific resonator and waveguide modes and may also be calculated from the mode profiles using nonorthogonal coupled mode theory (Fan et al., 2003; Haus and Huang, 1991). In order to remove the explicit time dependence from the total Hamiltonian, we can write the Hamiltonian operator in a frame rotating at the pump frequency (defined by the photon number operator). For a single pump laser, this is achieved by applying a unitary transformation $\mathcal{U} = \exp(-\mathrm{i}\omega_\mathrm{d} a_\mathrm{p}^\dagger a_\mathrm{p} t)$. The transformed Hamiltonian is obtained from the relation $\tilde{\mathcal{H}} = \mathcal{U}^\dagger \mathcal{H} \mathcal{U} + \mathrm{i}\hbar(\partial_t \mathcal{U})\mathcal{U}^\dagger$. When applying this transformation, each term in the Hamiltonian is multiplied by exponentials of the number operator, since these operators commute, they can be easily evaluated using the Baker–Campbell–Hausdorff formulas. For instance, the rotating drive reads

$$\begin{aligned} \frac{\mathcal{U}^\dagger \mathcal{H}^{(\mathrm{drive})} \mathcal{U}}{i\hbar E_p} &= \exp(\mathrm{i}\omega_\mathrm{p} a_\mathrm{p} a_\mathrm{p}^\dagger t)\left[a_\mathrm{p}^\dagger e^{-i\omega_\mathrm{p} t} + \mathrm{h.c.}\right] \exp(-\mathrm{i}\omega_\mathrm{p} a_\mathrm{p}^\dagger a_\mathrm{p} t) \\ &= a_\mathrm{p}^\dagger + a_\mathrm{p}, \end{aligned} \tag{90}$$

where the commutator (77a) was used. Therefore, the corresponding rotating wave Hamiltonian is written as,

$$\tilde{\mathcal{H}}/\hbar = \Omega b^\dagger b - \sum_n \Delta_n a_n^\dagger a_\mathrm{p} + iE_\mathrm{p}(a_\mathrm{p}^\dagger + a_\mathrm{p}) + \mathcal{H}^{(\mathrm{OM})}/\hbar, \tag{91}$$

where we introduced the pump-cavity detuning, $\Delta_\mathrm{p} = \delta_{n\mathrm{p}}\omega_\mathrm{p} - \omega_n$. Note that our rotating frame choice is not unique. For instance, when there is a single optical mode, our choice is often the most convenient since the corresponding rotating cavity mode operator will be a slowly varying one. However, when there are multiple optical modes under a single pump, it is often more convenient to introduce a rotating wave transformation to each optical field operator. The driven Heisenberg equations of motion may be obtained following steps analogous to previously described.

As we stated in the beginning of this section, the addition of an input channel to our resonator must be accompanied by a loss channel, which shall be treated with the formalism of open quantum systems (Agarwal, 2013; Bowen and Milburn, 2015; Breuer and Petruccione, 2002; Gardiner and Collett, 1985). We restrict ourselves here to writing the resulting quantum Langevin equations,

$$\frac{\partial}{\partial t} a_n = \left(\mathrm{i}\Delta_\mathrm{p} - \frac{\kappa}{2}\right) a_n - \mathrm{i}\sum_m \left(g_{n,m} b + g_{m,n}^* b^\dagger\right) a_m + \delta_{n\mathrm{p}} E_\mathrm{p} + \sqrt{\kappa} a_n^{(\mathrm{in})} \tag{92a}$$

$$\frac{\partial}{\partial t} b = \left(-\mathrm{i}\Omega - \frac{\Gamma}{2}\right) b - \mathrm{i}\sum_{n,m} \left(g_{n,m}^* a_m^\dagger a_n\right) + \sqrt{\Gamma} b^{(\mathrm{in})}, \tag{92b}$$

where κ_i is the intrinsic optical energy loss rates (such that $\kappa = \kappa_i + \kappa_e$), and Γ represents the mechanical mode energy loss rate.[l] Rigorously, the mechanical dissipation term is proportional to the phonon momentum quadrature $(b - b^\dagger)$, and we disregard the $b^\dagger$ term assuming a RWA with $\Omega \gg \Gamma$, a condition that is satisfied in common experimental conditions. The operators $a_n^{(\mathrm{in})}$, $b^{(\mathrm{in})}$ appearing alongside the dissipation terms are fluctuations due to noise entering the system from the bath. Due to the high optical mode frequencies, we can assume $a_n^{(\mathrm{in})}$ to be vacuum noise, with a zero mean value and nonzero autocorrelations that in the Markovian limit[m] are given by,

[l] When we write the term $\sqrt{\kappa} a_n^{(\mathrm{in})}$ for the fluctuation, we are implicitly stating κ is the total rate at which photons are escaping the resonator (e.g., coupling loss, radiation, absorption).

[m] The Markovian approximation is based on the assumption that the bath correlation functions decay on a time-scale much shorter than system correlation functions, leading to delta function autocorrelations (Bowen and Milburn, 2015).

$$\langle a_n^{(\mathrm{in})}(t) a_n^{(\mathrm{in})}(t')^\dagger \rangle = \delta(t - t'), \tag{93a}$$

$$\langle a_n^{(\mathrm{in})}(t)^\dagger a_n^{(\mathrm{in})}(t') \rangle = 0, \tag{93b}$$

$$\langle b_{(\mathrm{in})}(t) b_{(\mathrm{in})}(t') \rangle = (n_{\mathrm{th}} + 1)\delta(t - t'), \tag{93c}$$

$$\langle b^{(\mathrm{in})}(t)^\dagger b^{(\mathrm{in})}(t') \rangle = n_{\mathrm{th}}\delta(t - t'), \tag{93d}$$

where $n_{\mathrm{th}} = 1/[\exp(\hbar\Omega/k_B T) - 1]$ is the mechanical mode thermal occupation, which at room temperatures (large occupation) is approximately given by $n_{\mathrm{th}} \approx k_B T/\hbar\Omega$. The asymmetry in these autocorrelation functions already reveal that the photon and phonon creation/annihilation processes are not symmetric at low occupation numbers. Indeed, this property has been explored to identify the quantum regime of the optomechanical interaction (Safavi-Naeini et al., 2012).

To obtain the output field escaping the resonator, e.g., to the bus waveguide or coupling mirrors, a input-output relation is necessary, which may be derived by applying time-reversal symmetry in the quantum Langevin formalism, where it is assumed that the input channel feeding the resonator is the same that take energy away from it. In the classical realm, the field boundary conditions associated with time-reversal arguments also lead to similar input-output relations (Haus, 1984), which can be viewed as boundary conditions relating internal and external modes. For evanescently coupled traveling wave resonators or single-port Fabry–Perot resonators, the output fields is given by $a^{(\mathrm{out})} = (\alpha + a^{(\mathrm{in})}) - \sqrt{\kappa_e} a$, where α accounts for the coherent classical drive contribution (E_{p}) (Aspelmeyer et al., 2014; Bowen and Milburn, 2015). Since the vacuum noise has zero mean value ($\langle a^{(\mathrm{in})} \rangle = 0$), the expectation value for the output field recovers the classically expected relation $\langle a^{(\mathrm{out})} \rangle = \alpha - \sqrt{\kappa_e}\langle a \rangle$.

5.3 Typical settings

The set of equations (92) and the autocorrelation relations (93) can be used to describe an extensive set of experiments on resonator-enhanced Brillouin scattering. From single-mode optomechanical cooling and amplification (Chan et al., 2011; Gröblacher et al., 2009; Kippenberg et al., 2005) to Brillouin cooling and lasing (Grudinin et al., 2009, 2010; Otterstrom et al., 2018; Tomes and Carmon, 2009) or cascaded Brillouin lasing dynamics (Behunin et al., 2018). Specific settings and applications of these equations will be given throughout the book chapters.

5.3.1 Single-mode interaction

To gain an insight into the interaction predicted by Eq. (92), we inspect its behavior. In the case of a single optical mode optomechanical interaction, where we can disregard the $m \neq n$ terms,

$$\frac{\partial}{\partial t}a = \left[\mathrm{i}\Delta_{\mathrm{p}} - \frac{\kappa}{2} - \mathrm{i}g(b + b^{\dagger})\right]a + E_{\mathrm{p}}, \tag{94a}$$

$$\frac{\partial}{\partial t}b = \left(-\mathrm{i}\Omega - \frac{\Gamma}{2}\right)b - \mathrm{i}\sum_{n} g a^{\dagger} a, \tag{94b}$$

where we also applied the phase rotation to the eigenmodes to ensure g is real-valued, i.e., $g = g^*$. It is clear from Eq. (94) that the effect of the mechanical mode is to cause an oscillation (at the mechanical frequency) of the optical mode frequency ($g(b + b^{\dagger})$). One interesting feature of Eq. (94) is that both photon and phonon numbers are individually conserved, in the absence of both optical pump and dissipation:

$$\frac{\partial}{\partial t}a^{\dagger}a = a^{\dagger}\frac{\partial}{\partial t}a + \frac{\partial a^{\dagger}}{\partial t}a = 0, \tag{95a}$$

$$\frac{\partial}{\partial t}b^{\dagger}b = b^{\dagger}\frac{\partial}{\partial t}b + \frac{\partial b^{\dagger}}{\partial t}b = 0. \tag{95b}$$

This conclusion may sound contradictory for those familiar with cavity optomechanics experiments, where amplification (phonon creation) and cooling (phonon annihilation) are known to occur (Aspelmeyer et al., 2014). This is the case because we ignored the optical pump term that will translate these pure cavity frequency fluctuations to changes in the photon number, causing thus a back-action force on the mechanical mode ($ga^{\dagger}a$); Fig. 7 illustrates this concept. Despite the various similarities to FBS in waveguides (Section 4.2), the emergence of such back-action force in cavities is a striking difference. In contrast to the typical waveguide, the cavity resonance introduces a strong dispersion that underlies the back-action force.

Another key point in Eq. (94) is the optical loss rate (κ). When the pump is properly detuned from the cavity resonance, the symmetry between the competing phonon creation and annihilation processes is broken and the back-action force may amplify or cool the mechanical mode. Such effects are illustrated in Fig. 7A and B. As the mechanical frequency increases beyond the optical linewidth ($\Omega > \kappa/2$), the so-called resolved sideband limit is reached, and the phonon creation/annihilation asymmetry further increases. In particular, when the pump-cavity detuning matches the

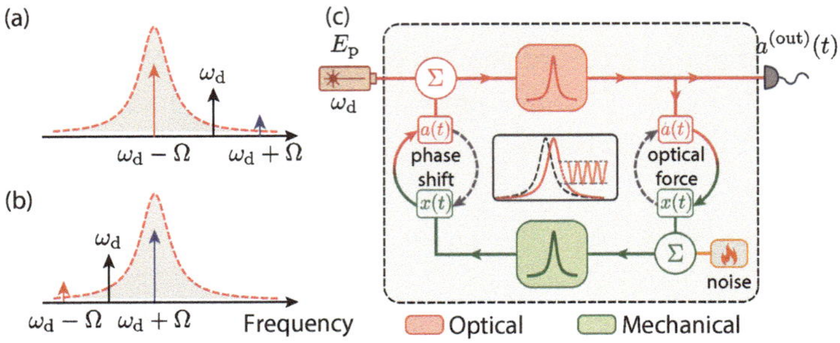

Fig. 7 Single-mode optomechanical interaction. The pump laser may be detuned either blue (A) or red (B) relative to the optical resonance to favor amplification (A) or cooling (B) of mechanical modes. (c) Representation of the optomechanical feedback. The mechanical degree of freedom (represented by $x(t)$) is in contact with the thermal noise bath, but when a continuous-wave laser excites the optical field within the optical resonator, the optical phase is imparted by the mechanical motion and converted to fluctuations on the optical energy. Due to radiation-pressure forces, the mechanical oscillator experiences a feedback (back-action) force that modify its dynamics.

mechanical frequency ($\Delta_p = \pm\Omega$), the optomechanical interaction is enhanced and the cooling and amplification processes efficiency are maximum. A more in-depth analysis of the single-mode interaction will be given in Chapter 3.

5.3.2 Two-mode interaction

The efficiency of the resolved sideband regime to reach phonon cooling (amplification) relies on the optical cavity linewidth filtering of the competing amplification (cooling) process. Such a filtering process, however, also prevents the pump from efficiently populating the driven cavity mode. In practice, this means that a strong pump laser is necessary to reach a large photon occupation of the driven cavity mode. In the context of cavity optomechanics, an alternative to overcome this pump efficiency penalty is to employ two optical cavity modes that are spaced apart by the mechanical frequency. This approach preserves the selectivity to either phonon creation or annihilation processes, while allowing for a resonant pump laser. In the context of intermodal FBS or intramodal backward Brillouin scattering, the two-mode interaction is ubiquitous. Indeed, the early experimental demonstrations of this effect in microresonators were feasible due to the two-mode interaction. Considering two optical modes in Eq. (80), $\omega_1 = \omega_p$ for the pump wave and $\omega_2 = \omega_S$ for the Stokes wave, such that $\omega_p = \omega_S + \Omega$. The Heisenberg equations are then given by

$$\frac{\partial}{\partial t}a_{\mathrm{p}} = -\mathrm{i}[\omega_{\mathrm{p}} + g_{\mathrm{p,p}}(b + b^{\dagger})]a_{\mathrm{p}} - \frac{\kappa_{\mathrm{p}}}{2}a_{\mathrm{p}} - \mathrm{i}(g_{\mathrm{p,S}}b + g^{*}_{\mathrm{S,p}}b^{\dagger})a_{\mathrm{S}} + E_{\mathrm{p}}e^{-\mathrm{i}\omega_{\mathrm{d}}t}, \tag{96a}$$

$$\frac{\partial}{\partial t}a_{\mathrm{S}} = -\mathrm{i}[\omega_{\mathrm{S}} + g_{\mathrm{S,S}}(b + b^{\dagger})]a_{\mathrm{S}} - \frac{\kappa_{\mathrm{S}}}{2}a_{\mathrm{S}} - \mathrm{i}(g_{\mathrm{S,p}}b + g^{*}_{\mathrm{p,S}}b^{\dagger})a_{\mathrm{p}}, \tag{96b}$$

$$\frac{\partial}{\partial t}b = (-\mathrm{i}\Omega - \Gamma/2)b - \mathrm{i}(g^{*}_{\mathrm{p,p}}a^{\dagger}_{\mathrm{p}}a_{\mathrm{p}} + g^{*}_{\mathrm{S,S}}a^{\dagger}_{\mathrm{S}}a_{\mathrm{S}}) - \mathrm{i}(g^{*}_{\mathrm{S,p}}a^{\dagger}_{\mathrm{p}}a_{\mathrm{S}} + g^{*}_{\mathrm{p,S}}a_{\mathrm{p}}a^{\dagger}_{\mathrm{S}}). \tag{96c}$$

We chose to not derive these from the rotating wave Hamiltonian because this is one example (as discussed in Section 5.2.1) where it would be more convenient to introduce a rotating frame for each optical mode. In the nonrotating frame above, we can clearly infer the oscillation frequencies of the each term in Eq. (96). As illustrated in Fig. 8A, depending on the ratio between optical mode linewidths ($\kappa_{(\mathrm{p,S})}$) and mechanical frequencies we may disregard the nonresonant terms under the RWA approximation. For instance, the term $g_{\mathrm{p,S}}{}^{*}b^{\dagger}a_{\mathrm{S}}$ in Eq. (96a) oscillates at $\omega = \omega_{\mathrm{S}} - \Omega$, clearly not resonant with the pump mode—this term is not illustrated in Fig. 8A. A similar argument applies toward the self-interaction terms. The mechanical mode is also driven by nonresonant terms and may be neglected in most experimental settings. However, it is worth mentioning that in some experimental conditions, intermodal forward or backward scattering may occur at very low frequencies

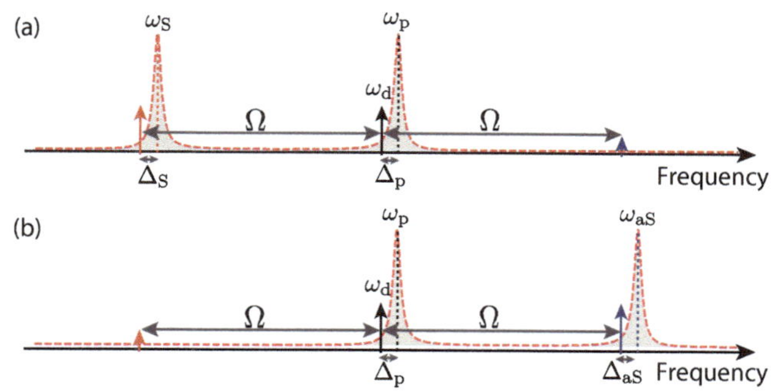

Fig. 8 Two-mode cavity Brillouin interaction. In the two-mode interaction, one may choose whether the Stokes (ω_{S}) (A) or anti-Stokes (ω_{aS}) (B) scattered wave will be resonant with the second cavity mode. This allows not only resonant enhancement of the chosen scattering process, but also to suppress the competing scattering process. As a result, efficient amplification or cooling of the mechanical wave may be achieved. Nevertheless, it still requires careful device engineering (Li et al., 2012) or optical tuning mechanisms to enable the ensure the resonance condition, as indicated by the relevant detuning parameters $\Delta_{(\mathrm{S,aS})}$

(He et al., 2020) and one must carefully evaluate the appropriateness of the RWA. Therefore, under such a strict RWA, the most common form of Eqs. (96) is given by,

$$\frac{\partial}{\partial t}a_{\mathrm{P}} = \left(-\mathrm{i}\omega_{\mathrm{P}} - \frac{\kappa_{\mathrm{P}}}{2}\right)a_{\mathrm{P}} - \mathrm{i}g_{\mathrm{P,S}}ba_{\mathrm{S}} + E_{\mathrm{P}}e^{-\mathrm{i}\omega_{\mathrm{d}}}, \tag{97a}$$

$$\frac{\partial}{\partial t}a_{\mathrm{S}} = \left(-\mathrm{i}\omega_{\mathrm{S}} - \frac{\kappa_{\mathrm{S}}}{2}\right)a_{\mathrm{S}} - \mathrm{i}g^{*}_{\mathrm{P,S}}b^{\dagger}a_{\mathrm{P}}, \tag{97b}$$

$$\frac{\partial}{\partial t}b = \left(-\mathrm{i}\Omega - \frac{\Gamma}{2}b\right) - \mathrm{i}g^{*}_{\mathrm{P,S}}a_{\mathrm{P}}a^{\dagger}_{\mathrm{S}}. \tag{97c}$$

In contrast to the single-mode dynamics, the previous equations predict that (in the absence of pump and dissipation) the photon and phonon numbers are not individually conserved:

$$\frac{\partial}{\partial t}a^{\dagger}_{\mathrm{P}}a_{\mathrm{P}} = -\mathrm{i}g_{\mathrm{P,S}}ba_{\mathrm{S}}a_{\mathrm{P}} + \text{h.c.}, \tag{98a}$$

$$\frac{\partial}{\partial t}a^{\dagger}_{\mathrm{S}}a_{\mathrm{S}} = \mathrm{i}g_{\mathrm{P,S}}ba_{\mathrm{S}}a_{\mathrm{P}} + \text{h.c.}, \tag{98b}$$

$$\frac{\partial}{\partial t}b^{\dagger}b = \mathrm{i}g_{\mathrm{P,S}}ba_{\mathrm{S}}a_{\mathrm{P}} + \text{h.c.}. \tag{98c}$$

Instead, pump photons are annihilated while a Stokes photon and a phonon are created. Further discussion on some of the characteristics of the two-mode dynamics will be given in Chapter 3.

6. Alternative approaches

In a field marking its centenary, there have naturally been quite a number of other approaches to modeling Brillouin scattering developed over the years. In this section, we examine a few of these and their relationship to the treatment just presented.

6.1 The classical field approach

We note here that the dynamical equations (69) can be derived entirely classically. This is the approach taken in the textbooks of Boyd (2008) and Agrawal (2012), who derived the equations for scalar density waves (see Section 6.2), and the extension to the full vector problem can be found in Wolff et al. (2015b) and Rakich et al. (2012). The method starts from the linear equations of elasticity (1) and Maxwell's equations (15), expands in eigenmodes of the waveguide, and then introduces coupling terms that describe the opto-acoustic interaction. This approach is formally equivalent to the approach given in Section 3, except that the quantities appearing in the dynamical equations are envelope functions rather than operators.

A full derivation of the classical approach can be found in Wolff et al. (2021) and given the similarity to the Hamiltonian treatment we do not repeat it here.

6.2 Traditional density-wave approach to backwards Brillouin scattering

6.2.1 Background to the scalar picture

The coupled mode treatment outlined above (whether obtained from a Hamiltonian or conventional picture) is general, flexible and extensible, and is well suited to describe Brillouin interactions in modern nanoscale integrated waveguides which offer strong field confinement and frequently use materials which are anisotropic, at least for the elastic field.

However, Brillouin scattering in bulk and traditional optical fiber systems have been successfully described for many years by a simpler well-known scalar theory (Boyd, 2008). We refer the reader to that reference for the details of the theory. Here we show how this simpler scalar theory can be obtained from the opto-elastic coupled mode theory just presented and so provide a point of connection for comparing results and quantities in the two pictures, and demonstrating that in the relevant regime, the full vector treatment does coincide with the well known but less general scalar picture.

In the theory described above, the elastic waves are described in terms of the displacement field $\boldsymbol{U}(\boldsymbol{r}, t)$, which in the classical regime, obeys the elastic wave Eq. (5). In the older scalar treatment, the primary dynamical variable is instead the *density variation* $\delta\rho(\boldsymbol{r}, t)$. In particular, this has been the traditional way of handling longitudinal backward SBS in fibers (Kobyakov et al., 2010). We will first state the core elements of this theory before making the connection to the vector treatment that has been our primary focus in this chapter.

6.2.2 Elements of the scalar theory

We write the density of the waveguide medium as

$$\rho(\boldsymbol{r}, t) = \rho_0(\boldsymbol{r}) + \delta\rho(\boldsymbol{r}, t), \tag{99}$$

where $\rho_0(\boldsymbol{r})$ is the constant equilibrium density distribution which defines the elastic waveguide.[n] The elastic wave is described in terms of the density variation $\delta\rho$, which satisfies the scalar wave equation (Boyd, 2008):

$$\frac{\partial^2 \delta\rho}{\partial t^2} - \Gamma' \nabla^2 \frac{\partial \delta\rho}{\partial t} - v_l^2 \nabla^2 \delta\rho = \nabla \cdot \boldsymbol{f}^{\mathrm{el}}. \tag{100}$$

Here Γ' is a damping coefficient and $v_l = \sqrt{M/\rho_0}$ is the speed of (compressive) sound linked to the longitudinal[o] modulus M. Finally, $\boldsymbol{f}^{\mathrm{el}}$ is the body force density for the electrostrictive interaction driven by the interfering pump and Stokes waves.[p] This force is the gradient of the *electrostrictive pressure*:

$$\boldsymbol{f}^{\mathrm{el}} = \nabla p^{\mathrm{el}} = \nabla \left(-\frac{\epsilon_0}{2} \gamma_e \langle |\boldsymbol{E}|^2(\boldsymbol{r}, t) \rangle \right), \tag{101}$$

where the angle brackets reflect the fact that the elastic field can only respond to the time-averaged electric field. The strength of the opto-elastic interaction is characterized by the *electrostrictive constant* (Landau and Lifshitz, 1986)

$$\gamma_e = \rho \frac{\partial \epsilon}{\partial \rho}, \tag{102}$$

where ϵ is as usual the relative dielectric constant, now treated as isotropic.

The corresponding influence of the opto-elastic coupling on the electromagnetic field enters as a contribution to the nonlinear polarization

$$\begin{aligned} \boldsymbol{P}^{(\mathrm{nl})}(\boldsymbol{r}, t) &= \epsilon_0 \Delta\epsilon^{(\mathrm{nl})} \boldsymbol{E} \\ &= \epsilon_0 \frac{\partial \epsilon}{\partial \rho} \delta\rho(\boldsymbol{r}, t) \boldsymbol{E} \\ &= \epsilon_0 \gamma_e \frac{\delta\rho(\boldsymbol{r}, t)}{\rho_0(\boldsymbol{r})} \boldsymbol{E}, \end{aligned} \tag{103}$$

in the optical wave equation.

As in the full theory, one can expand the density variation $\delta\rho$ in terms of propagating modes with frequency Ω and propagation constant q. Substituting Eq. (101) for the body force density and inserting scalar

[n] We note that some approaches to the fiber problem treat the optical fiber as having a uniform density ρ_0 throughout the core and cladding, while others specifically account for the density variation due to the Ge or other doping in the core.

[o] For Brillouin scattering in fluids, the appropriate velocity is $v_l = \sqrt{K/\rho_0}$ where K is the bulk modulus.

[p] The theory does not include the moving boundary/radiation pressure effects which are negligible in the systems to which this theory is typically applied.

expansions for the electromagnetic field, one can compute an approximation to the gain. A full derivation of this result can be found in the appendix of Kobyakov et al. (2010), as well as in chapter 9 of Boyd (2008), and we simply quote the result here: for a waveguide with a uniform density cross section ρ_0, the scalar gain in units of 1/m/W is

$$G_{\text{scalar}} = \frac{\gamma_e^2 \omega^2}{n_{\text{eff}} v_\ell c^3 \rho_0 \Gamma} \frac{1}{A_{\text{oa}}}, \tag{104}$$

where n_{eff} is the effective index of the waveguide, $\Gamma = q^2\Gamma' \approx 4n_{\text{eff}}^2\omega^2/c^2$ is the inverse of the phonon lifetime, and A_{oa} is the opto-acoustic coupling area, which can usually be taken to be equal to the effective area of the waveguide mode.

6.2.3 Derivation from the full opto-elastic theory

Let us see how we can recover this description from the full elastic picture already presented. The key element is that the change in density only reflects compression of the medium and ignores any shear contribution. Moreover, the density change not even distinguish between isotropic and anisotropic compressions. Instead, within the limits of small deformations, the density variation is related only to the trace of the strain tensor[q]:

$$\frac{\delta\rho}{\rho_0} = -\text{Tr}[\overleftrightarrow{S}] = -(S_{xx} + S_{yy} + S_{zz}), \tag{105}$$

and in the absence of better information, we assume that all strain components are identical, i.e., that the material is subject to a *hydrostatic* compression:

$$S_{ij} \approx -\frac{\delta\rho}{3\rho_0}\delta_{ij}. \tag{106}$$

We will also use the fact that for such a shear-free material, the complex stiffness tensor has the particularly simple form:

$$c_{ijkl} = \left[M + (\eta_{11} + 2\eta_{12})\frac{\partial}{\partial t}\right]\delta_{ij}\delta_{kl}, \tag{107}$$

where the *longitudinal modulus* $M = c_{11} + 2c_{12}$ and the corresponding viscosity terms have both been written in the compact Voigt notation

[q] This follows from the observation that the change in volume V and the density satisfy $\frac{\delta\rho}{\rho_0} = -\frac{\delta V}{V} = -\nabla \cdot \boldsymbol{U}$.

(see Appendix). This form ensures that the stress tensor will likewise be shear free, with diagonal elements that are identical and equal to the hydrostatic pressure p.

An equation of motion for $\delta\rho$ can be found either directly via the Hamiltonian using the Heisenberg equation (26), or simply by taking the divergence of Eq. (5):

$$\rho_0 \frac{\partial^2}{\partial t^2}\left(-\frac{\delta\rho}{\rho_0}\right) = \frac{\partial}{\partial r_i}\frac{\partial}{\partial r_j} T_{ij} + \frac{\partial f_i}{\partial r_i}\,. \tag{108}$$

Using the above hydrostatic results, this becomes

$$\begin{aligned}\frac{\partial^2 \delta\rho}{\partial t^2} &= -\frac{\partial}{\partial r_i}\frac{\partial}{\partial r_j}\left(M + (\eta_{11} + 2\eta_{12})\frac{\partial}{\partial t}\right)\delta_{ij}\delta_{kl}S_{kl} - \frac{\partial f_i}{\partial r_i}\\ &= -\rho_0\left(v_l^2 + \Gamma'\frac{\partial}{\partial t}\right)\frac{\partial}{\partial r_i}\frac{\partial}{\partial r_i}S_{kk} - \frac{\partial f_i}{\partial r_i}\\ &= \left(v_l^2 + \Gamma'\frac{\partial}{\partial t}\right)\nabla^2\delta\rho - \nabla\cdot\boldsymbol{f},\end{aligned} \tag{109}$$

where we have introduced the longitudinal phase velocity $v_l = \sqrt{M/\rho_0}$ and the damping factor $\Gamma' = (\eta_{11} + 2\eta_{12})/\rho_0$. Here, the body force density $\boldsymbol{f}$ represents the electrostrictive force induced by the optical fields. This equation will be equivalent to Eq. (100), if we can show that the body force term $-\nabla\cdot\boldsymbol{f}$ matches the divergence $\nabla\cdot\boldsymbol{f}^{\mathrm{el}}$ of the electrostrictive force $\boldsymbol{f}^{\mathrm{el}}$ defined in Eq. (101).

6.2.4 Evaluation of the force term

We now consider the driving term $-\nabla\cdot\boldsymbol{f}$ appearing in Eq. (109), deriving this term from the photoelastic interaction term in the Hamiltonian of the complete theory. We do so by noting that the force density $\boldsymbol{f}(\boldsymbol{r}, t)$ is equal to the time rate of change of the momentum density $\boldsymbol{\pi}(\boldsymbol{r}, t)$, and so using Eqs. (26) and (35a), we find the mth component of the force is

$$\begin{aligned}f_m(\boldsymbol{r}, t) &= \frac{\partial \pi_n}{\partial t} = \frac{1}{i\hbar}\left[\pi_m, \mathcal{H}^{(\mathrm{PE})}\right]\\ &= \frac{1}{i\hbar}\int \mathrm{d}^3 r\, \left[\pi_m(\boldsymbol{r}), \frac{1}{2\epsilon_0}D_i(\boldsymbol{r})p_{ijkl}S_{kl}D_j(\boldsymbol{r})\right].\end{aligned} \tag{110}$$

The commutator is evaluated using similar manipulations to those leading to Eq. (27b), which ultimately yields

$$f_m(\boldsymbol{r},t) = \frac{1}{2\epsilon_0} p_{ijkm} \frac{\partial}{\partial r_k} D_i(\boldsymbol{r}) D_j(\boldsymbol{r}), \tag{111}$$

so that

$$\begin{aligned} -\nabla \cdot \boldsymbol{f} &= -\frac{1}{2\epsilon_0} \frac{\partial}{\partial r_m} p_{ijkm} \frac{\partial}{\partial r_k} \langle D_i(\boldsymbol{r},t) D_j(\boldsymbol{r},t) \rangle \\ &= -\frac{\epsilon_0}{2} p_{ijkm}\, \epsilon(\boldsymbol{r})^2 \frac{\partial}{\partial r_m} \frac{\partial}{\partial r_k} \langle E_i(\boldsymbol{r},t) E_j(\boldsymbol{r},t) \rangle, \end{aligned} \tag{112}$$

where we used $\boldsymbol{D} = \epsilon_0 \epsilon \boldsymbol{E}$, assumed p_{ijkl} and $\epsilon(\boldsymbol{r})$ are piecewise constant, and added the angle brackets to recognize that the elastic field can only respond to the time-averaged electromagnetic field.

6.2.5 Linearly polarized approximation

We now seek to relate Eq. (112) to the form in Eq. (101). In standard fiber, it is reasonable to treat the optical field as linearly polarized such that $\boldsymbol{E}(\boldsymbol{r}) = E_x(\boldsymbol{r})\hat{\boldsymbol{x}}$ giving

$$-\nabla \cdot \boldsymbol{f} = -\frac{\epsilon_0}{2} p_{xxkm}\, \epsilon(\boldsymbol{r})^2 \frac{\partial}{\partial r_m} \frac{\partial}{\partial r_k} \langle E_x(\boldsymbol{r}) E_x(\boldsymbol{r}) \rangle. \tag{113}$$

In isotropic media, the Pockel's tensor has a particularly simple form with only the p_{11} and p_{12} terms surviving (see Appendix). Therefore

$$\begin{aligned} -\nabla \cdot \boldsymbol{f} &= -\frac{\epsilon_0 \epsilon(\boldsymbol{r})^2}{2} \left[p_{11} \frac{\partial^2}{\partial x^2} + p_{12} \left(\frac{\partial^2}{\partial y^2} + \frac{\partial^2}{\partial z^2} \right) \right] \langle E_x(\boldsymbol{r},t)^2 \rangle \\ &= -\frac{\epsilon_0 \epsilon(\boldsymbol{r})^2}{2} \left[p_{12} \nabla^2 + (p_{11} - p_{12}) \frac{\partial^2}{\partial x^2} \right] \langle E_x(\boldsymbol{r},t)^2 \rangle \\ &\approx -\frac{\epsilon_0 \epsilon(\boldsymbol{r})^2}{2} p_{12} \nabla^2 \langle E_x(\boldsymbol{r},t) \rangle^2, \end{aligned} \tag{114}$$

where we have dropped the second term in square brackets in the middle line on the basis that fiber modes are weakly guided and transverse derivatives small.

Using Eq. (A.12), the variation in the (scalar) relative dielectric constant is given by

$$\delta\epsilon = \frac{1}{3}\delta\epsilon_{ii} \tag{115}$$

$$= -\frac{1}{3}\epsilon^2\left(p_{xxkl} + p_{yykl} + p_{zzkl}\right)S_{kl}$$

$$= -\frac{1}{3}\epsilon^2(2p_{12} + p_{11})\mathrm{Tr}[\overleftrightarrow{S}] \tag{116}$$

$$= \epsilon^2\left(p_{12} + \frac{(p_{11} - p_{12})}{3}\right)\frac{\delta\rho}{\rho_0}.$$

From this relation and Eq. (102), we can identify the electrostrictive constant as

$$\gamma_e = \epsilon^2 p_{12}\left(1 + \frac{(p_{11} - p_{12})}{3p_{12}}\right) \approx \epsilon^2 p_{12}. \tag{117}$$

Finally, we can now rewrite Eq. (114) as

$$\begin{aligned} -\nabla \cdot \boldsymbol{f} &= \nabla \cdot \left[\nabla\left(-\frac{\epsilon_0}{2}\gamma_e \nabla^2 \langle E_x(\boldsymbol{r}, t)\rangle^2\right)\right] \\ &= \nabla \cdot \boldsymbol{f}^{\mathrm{el}}, \end{aligned} \tag{118}$$

in agreement with the scalar theory's Eq. (101).

6.3 The Lagrangian approach and acoustic subordinate equations

Apart from a full ab-initio treatment of the equations of motion (which tends to be very challenging and in many situations a bit wasteful), the simplified density approach for fibers, and derivations of the CMEs directly from Maxwell's equations and the equations of linear elasticity, or via a Hamiltonian picture as above, there is also the possibility of obtaining the dynamical equations from a Lagrangian picture. This is in many ways very similar to the Hamiltonian picture, from which it can be obtained in the lossless case via a Legendre transformation. The resulting total Lagrangian that minimizes the total action of the nonlinearly coupled optical and acoustic systems (Laude and Beugnot, 2015) is

$$\mathcal{L}^{(\mathrm{tot})} = \int \mathrm{d}^3 r \left[\underbrace{\frac{\rho|\dot{\boldsymbol{U}}|^2 - \overleftrightarrow{S} : \overleftrightarrow{c} : \overleftrightarrow{S}}{2}}_{=\mathcal{L}^{(\mathrm{mech})}} + \underbrace{\frac{\boldsymbol{E}\cdot\boldsymbol{D} - \boldsymbol{H}\cdot\boldsymbol{B}}{2}}_{=\mathcal{L}^{(\mathrm{EM})}} + \mathcal{L}^{(\mathrm{OM})}\right], \tag{119}$$

where $\mathcal{L}^{(\mathrm{mech})}$, $\mathcal{L}^{(\mathrm{EM})}$, and $\mathcal{L}^{(\mathrm{OM})}$ are the mechanical, electromagnetic, and optomechanical Lagrange densities. Note how the Legendre transformation has changed the signs in front of the magnetic and stress–strain terms compared to the Hamiltonian approach. Since the only relevant optomechanical interaction processes involve the electric fields and displacement or strain, but not the magnetic or the velocity field, the interaction Lagrangian is the sum of the three relevant interaction Hamiltonians except for an overall minus sign:

$$\mathcal{L}^{(\mathrm{OM})} = -\mathcal{H}^{(\mathrm{ES})} - \mathcal{H}^{(\mathrm{MB})} - \mathcal{H}^{(\mathrm{RO})}. \tag{120}$$

From this, the equations of motion are obtained via the Euler–Lagrange equation rather than the Heisenberg equations of the Hamiltonian system. Upon a modal expansion ansatz and appropriate approximations (namely SVEA and RWA), the CMEs can be recovered.

If used to merely derive the conventional CMEs, the Lagrangian approach has no real advantage over the Hamiltonian one and the derivation of a quantum theory is more difficult. However, where it really shines is the systematic incorporation of loss via a classic dissipation function (Landau and Lifshitz, 1986). This allows the introduction of linear and especially nonlinear lossy processes, apply consistent (e.g., inherently momentum conserving) approximations similar to how the RWA is applied in a Hamiltonian framework, and derive all resulting contributions to the CMEs.

Another example for leaving the strict modal picture mostly used for Brillouin theory is what for lack of a better term we might call the acoustic subordinate formulation. Here, the optical fields are still found as eigenmodes of the waveguide, but the mechanical deformation field is found as the solution to a (linear or nonlinear) response problem with the optical forces acting as the driving term. This approach is fundamentally well suited for dealing with open acoustic boundaries, higher order optomechanical nonlinearities and large-core waveguides where a modal expansion would require the contribution of many spectrally overlapping acoustic eigenmodes. This is paid for by a less elegant numerical calculation of the Brillouin gain by determining the sensitivity of the optical dispersion relation to the acoustic strain field by subsequent simulations of the optical problem including finite deformations (Laude and Beugnot, 2015) instead of a straightforward overlap integral.

7. Outlook and the limits of the formalism

Within this chapter we have presented the formalism of Brillouin scattering in as general a form as possible, while attempting not to become overly bogged down in detailed situations that might arise. Thus, in the form of Eqs. (55a) and (55b), we can compute the optical and acoustic fields for forward, backward, and intermodal Brillouin configurations for any given combination of input pulses in any number of modes.

There are several situations that the theory presented here will not cover. We outline these briefly here:

- *Interaction with other nonlinear effects and dispersion*: The presence of second- and third-order nonlinearities can significantly alter the Brillouin response, especially for situations where the optical intensity becomes large. Fortunately the extension to the theory presented here to include nonlinear effects is straightforward: one can include for example the Kerr nonlinearity by addition of the usual nonlinear terms into the optical equations (55a). A full set of dynamic equations including the Kerr effect can be found in Winful (2013). Dispersion and the Raman effect can also be incorporated in this way—we note, however, that most Brillouin experiments involve pulses with durations of ns or longer, and so these latter effects do not in general play a great role. As for the theory of SBS interacting with nonlinear loss (e.g., two-photon absorption or resulting free carrier absorption), we point the reader to Wolff et al. (2015a) in waveguides and Mirnaziry et al. (2017) in weakly coupled resonators.
- *Highly lossy waveguides*: Throughout this chapter we have treated loss perturbatively, where the underlying assumption is that the modes (both optical and acoustic) can be approximated by their lossless counterparts, and the loss can be added in later as a dissipative term in the CMEs. This will no longer apply in situations where the loss is extremely high, as might occur, for example, in plasmonic waveguides or resonators (Primo et al., 2020). Lossy modes involve fields that diverge in one direction in either space or time, and this leads to problems in evaluating the coupling terms in the CMEs. To properly model these situations it would be necessary to re-formulate these equations based on a dual space containing adjoint modes. An alternative is to use a direct solver along the lines of the "master/subordinate" model discussed in Section 6.3.

- *Ultra-short pulses*: We have assumed in our derivation of the coupled mode equations that the pulses are sufficiently long that higher order derivatives in time can be neglected, and that the RWA remains valid. This is not always the case, and in particular several experiments have demonstrated that Brillouin effects persist even down to the 100ps level (Merklein et al., 2018). The extension of the theory to shorter pulses (down to the ps regime) requires has been presented in Piotrowski et al. (2021).

Acknowledgments

C.G.P. and M.J.S. acknowledge support from the Australian Research Council Discovery Project scheme under award DP200101893. C.W. acknowledges funding from the Villum Fonden related to Grant No. 16498.

Appendix. Summary of elastic properties

In this Appendix, we summarize some commonly used properties of the linear and nonlinear elastic response tensors. For a comprehensive discussion including justifications for the relations listed here without proof, we refer to Auld (1990a).

A.1 General materials

For all materials, the strain and stress are both symmetric second-rank tensors. In other words, for all $i, j \epsilon \{x, y, z\}$,

$$S_{ij} = S_{ji} \tag{A.1a}$$

$$T_{ij} = T_{ji}. \tag{A.1b}$$

For all materials, the stiffness tensor satisfies a number of useful symmetries:

$$c_{ijkl} = c_{jikl} \tag{A.2a}$$

$$c_{ijkl} = c_{ijlk} \tag{A.2b}$$

$$c_{ijkl} = c_{klij}. \tag{A.2c}$$

A.2 Voigt notation

The symmetric nature of $\overleftrightarrow{S}$ and $\overleftrightarrow{T}$ means there are only six independent components to each tensor. The stress and strain are frequently written in compact *Voigt* notation with a single upper case index:

$$S_I = \begin{bmatrix} S_1 \\ S_2 \\ S_3 \\ S_4 \\ S_5 \\ S_6 \end{bmatrix} = \begin{bmatrix} S_{xx} \\ S_{yy} \\ S_{zz} \\ 2S_{yz} \\ 2S_{xz} \\ 2S_{xy} \end{bmatrix}, \qquad T_I = \begin{bmatrix} T_1 \\ T_2 \\ T_3 \\ T_4 \\ T_5 \\ T_6 \end{bmatrix} = \begin{bmatrix} T_{xx} \\ T_{yy} \\ T_{zz} \\ T_{yz} \\ T_{xz} \\ T_{xy} \end{bmatrix}. \tag{A.3}$$

(The factor of 2 in the final three elements of the strain vector is a choice that simplifies other relations.) It follows from these definitions and the symmetry properties of the stiffness c_{ijkl} that tensor may also be written in a compact two index form according to

$$c_{IJ} = c_{ijkl}, \tag{A.4}$$

where the subscripts I, J obey the same associations

$$\begin{array}{lll} 1 \rightarrow xx, & 2 \rightarrow yy, & 3 \rightarrow zz \\ 4 \rightarrow yz, zy, & 5 \rightarrow xz, zx, & 6 \rightarrow xy, yx \end{array}$$

A.3 Viscosity

Eq. (3) is the mechanical constitutive relation between the strain and the total Cauchy stress due to the stiffness $\overleftrightarrow{c}$ and the dynamic viscosity $\overleftrightarrow{\eta}$ of the material:

$$T_{ij} = c_{ijkl} S_{kl} + \eta_{ijkl} \frac{\partial S_{kl}}{\partial t}. \tag{A.5}$$

If we assume a time-harmonic ansatz with frequency Ω, this becomes:

$$T_{ij} = c_{ijkl} S_{kl} - \mathrm{i}\Omega \eta_{ijkl} S_{kl}, \tag{A.6}$$

which indicates that we can either treat the mechanical response as the superposition of an in-phase elastic stress and an out-of-phase viscous stress. Alternatively, we can also combine both and introduce a complex-valued stiffness:

$$\tilde{c}_{ijkl}(\Omega) = c_{ijkl} - \mathrm{i}\Omega \eta_{ijkl}, \tag{A.7}$$

in analogy how the electric response of a material can be either understood as combination of real permittivity and conductance or in terms of a complex permittivity. More relevant is that Eq. (A.7) implies that $\overleftrightarrow{c}$ and $\overleftrightarrow{\eta}$ have the same symmetry properties.

A.4 Nonlinear properties

The nonlinear local response to a deformation field will to lowest order generally take the form

$$(\Delta\beta)_{ij} = (\Delta\beta)_{ij}\xi_{ijkl}\partial_k u_l, \tag{A.8}$$

with some fourth-rank tensor ξ_{ijkl}. From the fact that (in the absence of magneto-optic effects) permittivities are always symmetric, we can already conclude the property

$$\xi_{ijkl} = \xi_{jikl}. \tag{A.9}$$

Next, we note that it is customary to decompose the total gradient of the deformation field into a symmetric part $S_{kl} = \frac{1}{2}(\partial_k u_l + \partial_l u_k)$, which is the strain, and an antisymmetric part $R_{kl} = \frac{1}{2}(\partial_k u_l - \partial_l u_k)$, which is the local rotation or twist field. Therefore, we decompose the tensor ξ_{ijkl} into one that is symmetric in the second index pair

$$p_{ijkl} = p_{ijlk}; \tag{A.10}$$

which therefore only interacts with the strain field and describes photoelasticity, and one tensor that is antisymmetric in the second index pair

$$r_{ijkl} = -r_{ijlk}, \tag{A.11}$$

which therefore only interacts with the rotation field and describes the roto-optic effect. The total change in the inverse permittivity is then:

$$(\Delta\beta)_{ij} = p_{ijkl}S_{kl} + r_{ijkl}R_{kl}. \tag{A.12}$$

Since the photoelastic tensor has the similar symmetry properties to the stiffness tensor, it is also common to write it in Voigt notation.

A.5 Isotropic materials

For most materials, the stiffness displays additional symmetries according to the crystal class of the medium. In this Appendix, we summarize these for the two most common cases of isotropic and cubic materials.

A.5.1 Linear properties

For the simplest case of isotropic materials, the stiffness has the form (Auld, 1990a)

$$c_{IJ} = \begin{bmatrix} c_{11} & c_{12} & c_{12} & 0 & 0 & 0 \\ c_{12} & c_{11} & c_{12} & 0 & 0 & 0 \\ c_{12} & c_{12} & c_{11} & 0 & 0 & 0 \\ 0 & 0 & 0 & c_{44} & 0 & 0 \\ 0 & 0 & 0 & 0 & c_{44} & 0 \\ 0 & 0 & 0 & 0 & 0 & c_{44} \end{bmatrix}, \quad \text{(A.13)}$$

with the constraint that $c_{44} = (c_{11} - c_{12})/2$. There are thus only two free parameters (tied to the phase velocities of longitudinal and shear waves). The properties of the medium can be specified by any two of the following constants: c_{11}, μ, λ, ν, E, K. Here the two *Lamé parameters* are defined as

$$\mu = c_{44}, \qquad \lambda = c_{12}. \quad \text{(A.14)}$$

These parameters can alternatively be expressed in terms of Young's modulus E and the Poisson ratio ν:

$$\mu = \frac{E}{2(1+\nu)}, \qquad \lambda = \frac{E\nu}{(1+\nu)(1-2\nu)}, \quad \text{(A.15)}$$

and other combinations can be found involving the bulk modulus K (the inverse of the bulk compressibility) which satisfies

$$K = \frac{E}{3(1-2\nu)}. \quad \text{(A.16)}$$

The stress–strain relation (Hooke's law) for an isotropic material can be written

$$\overleftrightarrow{T} = \lambda \mathrm{Tr}[\overleftrightarrow{S}]\,\overleftrightarrow{I} + 2\mu\overleftrightarrow{S} = \lambda\nabla\cdot\boldsymbol{U}\,\overleftrightarrow{I} + 2\mu\overleftrightarrow{S} \quad \text{(A.17)}$$

or

$$T_{ij} = \lambda\frac{\partial U_k}{\partial r_k}\delta_{ij} + 2\mu S_{ij}. \quad \text{(A.18)}$$

On inserting plane wave solutions into the elastic wave equation for uniform isotropic materials, one finds longitudinal and shear phase velocities of

$$v_l = \sqrt{\frac{\lambda + 2\mu}{\rho}} = \sqrt{\frac{c_{11}}{\rho}}, \tag{A.19}$$

$$v_s = \sqrt{\frac{\mu}{\rho}} = \sqrt{\frac{c_{44}}{\rho}}. \tag{A.20}$$

An equivalent expression for the longitudinal velocity as

$$v_l = \sqrt{\frac{3K(1-\nu)}{\rho(1+\nu)}}, \tag{A.21}$$

which tends to $v_l \to \sqrt{K/\rho}$ in gases, for which the Poisson ratio is $\nu = 0.5$.

A.5.2 Nonlinear properties

For isotropic materials, p_{ijkl} has precisely the same form as the stiffness:

$$\bar{p} = \begin{bmatrix} p_{11} & p_{12} & p_{12} & 0 & 0 & 0 \\ p_{12} & p_{11} & p_{12} & 0 & 0 & 0 \\ p_{12} & p_{12} & p_{11} & 0 & 0 & 0 \\ 0 & 0 & 0 & p_{44} & 0 & 0 \\ 0 & 0 & 0 & 0 & p_{44} & 0 \\ 0 & 0 & 0 & 0 & 0 & p_{44} \end{bmatrix}, \tag{A.22}$$

with $p_{44} = (p_{11} - p_{12})/2$. In liquids, $p_{11} = p_{12}$ and $p_{44} = 0$.

A.6 Cubic materials

For cubic materials, the form of the stiffness is identical to the isotropic case in Eq. (A.13) but the three values c_{11}, c_{12}, and c_{44} are now independent—the isotropic constraint $c_{44} = (c_{11} - c_{12})/2$ no longer holds.

References

Agarwal, G.S., 2013. Quantum Optics. Cambridge University Press, ISBN: 9781107006409. https://books.google.com.br/books?id=7KKw_XIYaioC.

Agrawal, G.P., 2012. Nonlinear Fiber Optics, fifth ed. Academic Press, p. 529.

Aspelmeyer, M., Kippenberg, T.J., Marquardt, F., 2014. Cavity optomechanics. Rev. Mod. Phys. 86 (4), 1391–1452. https://doi.org/10.1103/RevModPhys.86.1391.

Auld, B.A., 1990a. Acoustic Fields and Waves in Solids, second ed. 1 Krieger Publishing Co., ISBN: 9780898747829.

Auld, B.A., 1990b. Acoustic Fields and Waves in Solids, second ed. vol. 2 Krieger Publishing Co., ISBN: 9780898747829.

Behunin, R.O., Otterstrom, N.T., Rakich, P.T., Gundavarapu, S., Blumenthal, D.J., 2018. Fundamental noise dynamics in cascaded-order Brillouin lasers. Phys. Rev. A 98 (2), 1–17. https://doi.org/10.1103/PhysRevA.98.023832.

Bhat, N.A.R., Sipe, J.E., 2006. Hamiltonian treatment of the electromagnetic field in dispersive and absorptive structured media. Phys. Rev. A 73, 063808. https://doi.org/10.1103/PhysRevA.73.063808.

Born, M., Infeld, L., 1934. On the quantization of the new field equations I. Proc. R. Soc. Lond. A 147, 522–546.

Bowen, W.P., Milburn, G.J., 2015. Quantum Optomechanics. Taylor & Francis, ISBN: 9781482259155. https://books.google.com.br/books?id=xqlcrgEACAAJ.

Boyd, R.W., 2008. Nonlinear Optics, third ed. Academic Press.

Breuer, H.P., Petruccione, F., 2002. The Theory of Open Quantum Systems. Oxford University Press, ISBN: 9780198520634. https://books.google.com.br/books?id=0Yx5VzaMYm8C.

Chan, J., Alegre, T.P.M., Safavi-Naeini, A.H., Hill, J.T., Krause, A., Gröblacher, S., Aspelmeyer, M., Painter, O., 2011. Laser cooling of a nanomechanical oscillator into its quantum ground state. Nature 478 (7367), 89–92. https://doi.org/10.1038/nature10461.

Clerk, A.A., Devoret, M.H., Girvin, S.M., Marquardt, F., Schoelkopf, R.J., 2010. Introduction to quantum noise, measurement, and amplification. Rev. Mod. Phys. 82, 1155–1208. https://doi.org/10.1103/RevModPhys.82.1155.

Drummond, P.D., Corney, J.F., 2001. Quantum noise in optical fibers I Stochastic equations. J. Opt. Soc. Am. B 18 (2), 139. https://doi.org/10.1364/josab.18.000139.

Espinel, Y.A.V., Santos, F.G.S., Luiz, G.O., Alegre, T.P.M., Wiederhecker, G.S., 2017. Brillouin optomechanics in coupled silicon microcavities. Sci. Rep. 7 (1), 43423. https://doi.org/10.1038/srep43423.

Fan, S., Suh, W., Joannopoulos, J.D., 2003. Temporal coupled-mode theory for the fano resonance in optical resonators. J. Opt. Soc. Am. A 20 (3), 569–572. https://doi.org/10.1364/JOSAA.20.000569. http://opg.optica.org/josaa/abstract.cfm?URI=josaa-20-3-569.

Florez, O., Jarschel, P.F., Espinel, Y.A.V., Cordeiro, C.M.B., Alegre, T.P.M., Wiederhecker, G.S., Dainese, P., 2016. Brillouin scattering self-cancellation. Nat. Commun. 7, 11759. https://doi.org/10.1038/ncomms11759.

Gardiner, C.W., Collett, M.J., 1985. Input and output in damped quantum systems: quantum stochastic differential equations and the master equation. Phys. Rev. A 31 (6), 3761–3774. https://doi.org/10.1103/PhysRevA.31.3761.

Gröblacher, S., Hertzberg, J.B., Vanner, M.R., Cole, G.D., Gigan, S., Schwab, K.C., Aspelmeyer, M., 2009. Demonstration of an ultracold micro-optomechanical oscillator in a cryogenic cavity. Nat. Phys. 5 (7), 485–488. https://doi.org/10.1038/nphys1301.

Grudinin, I.S., Matsko, A.B., Maleki, L., 2009. Brillouin lasing with a CaF_2 whispering gallery mode resonator. Phys. Rev. Lett. 102 (4), 043902. https://doi.org/10.1103/PhysRevLett.102.043902.

Grudinin, I.S., Lee, H., Painter, O., Vahala, K.J., 2010. Phonon laser action in a tunable two-level system. Phys. Rev. Lett. 104 (8), 083901. https://doi.org/10.1103/PhysRevLett.104.083901.

Haus, H.A., 1984. Waves and Fields in Optoelectronics. Prentice-Hall Series in Solid State Physical Electronics, Prentice-Hall, ISBN: 9780139460531. https://books.google.com.br/books?id=AQUpAQAAMAAJ.

Haus, H.A., Huang, W., 1991. Coupled-mode theory. Proc. IEEE 79 (10), 1505–1518. https://doi.org/10.1109/5.104225.

He, X., Harris, G.I., Baker, C.G., Sawadsky, A., Sfendla, Y.L., Sachkou, Y.P., Forstner, S., Bowen, W.P., 2020. Strong optical coupling through superfluid Brillouin lasing. Nat. Phys. 16 (4), 417–421. https://doi.org/10.1038/s41567-020-0785-0.

Honda, Y., Yoshiki, W., Tetsumoto, T., Fujii, S., Furusawa, K., Sekine, N., Tanabe, T., 2018. Brillouin lasing in coupled silica toroid microcavities. Appl. Phys. Lett. 112 (20), 201105. https://doi.org/10.1063/1.5021062.

Joannopoulos, J.D., Johnson, S.G., Winn, J.N., Meade, R.D., 2011. Photonic Crystals: Molding the Flow of Light. Princeton University Press.

Johnson, S.G., Joannopoulos, J.D., 2001. Block-iterative frequency-domain methods for Maxwell's equations in a planewave basis. Opt. Express 8 (3), 173–190. https://doi.org/10.1364/OE.8.000173. http://www.osapublishing.org/oe/abstract.cfm?URI=oe-8-3-173.

Johnson, S., Ibanescu, M., Skorobogatiy, M., Weisberg, O., Joannopoulos, J., Fink, Y., 2002. Perturbation theory for Maxwell's equations with shifting material boundaries. Phys. Rev. E 65 (6), 066611. https://doi.org/10.1103/PhysRevE.65.066611.

Kippenberg, T.J., Rokhsari, H., Carmon, T., Scherer, A., Vahala, K.J., 2005. Analysis of radiation-pressure induced mechanical oscillation of an optical microcavity. Phys. Rev. Lett. 95 (3), 033901. https://doi.org/10.1103/PhysRevLett.95.033901.

Kobyakov, A., Sauer, M., Chowdhury, D., 2010. Stimulated Brillouin scattering in optical fibers. Adv. Opt. Photonics 2 (1), 1. https://doi.org/10.1364/AOP.2.000001. http://www.opticsinfobase.org/abstract.cfm?URI=aop-2-1-1.

Kornreich, P., Callen, H., Gundjian, A., 1967. "Current striction"—a mechanism of electrostriction in many-valley semiconductors. Phys. Rev. 161 (3), 815–822. https://doi.org/10.1103/PhysRev.161.815.

Laer, R.V., Kuyken, B., Thourhout, D.V., Baets, R., 2015. Interaction between light and highly confined hypersound in a silicon photonic nanowire. Nat. Photonics 9, 199–203. https://doi.org/10.1038/nphoton.2015.11.

Landau, L.D., Lifshitz, E.M., 1976. Mechanics, third ed. Pergamon Press, Oxford, UK, p.p. 170.

Landau, L.D., Lifshitz, E.M., 1986. Theory of Elasticity, third ed. Pergamon Press, Oxford, UK, p.p. 187.

Laude, V., Beugnot, J.-C., 2015. Lagrangian description of Brillouin scattering and electrostriction in nanoscale optical waveguides. New J. Phys. 17 (12), 125003. https://doi.org/10.1088/1367-2630/17/12/125003.

Li, J., Lee, H., Chen, T., Vahala, K.J., 2012. Characterization of a high coherence, Brillouin microcavity laser on silicon. Opt. Express 20 (18), 20170–20180. https://doi.org/10.1364/OE.20.020170. http://opg.optica.org/oe/abstract.cfm?URI=oe-20-18-20170.

Merklein, M., Stiller, B., Eggleton, B.J., 2018. Brillouin-based light storage and delay techniques. J. Opt. 20 (8), 083003. https://doi.org/10.1088/2040-8986/aad081.

Mirnaziry, S.R., Wolff, C., Steel, M.J., Morrison, B., Eggleton, B.J., Poulton, C.G., 2017. Lasing in ring resonators by stimulated Brillouin scattering in the presence of nonlinear loss. Opt. Express 25 (20), 23619–23633. https://doi.org/10.1364/OE.25.023619. http://opg.optica.org/oe/abstract.cfm?URI=oe-25-20-23619.

Nelson, D.F., Lax, M., 1970. New symmetry for acousto-optic scattering. Phys. Rev. Lett. 24 (8), 379–380. https://doi.org/10.1103/PhysRevLett.24.379.

Nieves, O.A., Arnold, M.D., Steel, M.J., Schmidt, M.K., Poulton, C.G., 2021a. Noise and pulse dynamics in backward stimulated Brillouin scattering. Opt. Express 29 (3), 3132. https://doi.org/10.1364/oe.414420.

Nieves, O.A., Arnold, M.D., Steel, M.J., Schmidt, M.K., Poulton, C.G., 2021b. Numerical simulation of noise in pulsed Brillouin scattering. J. Opt. Soc. Am. B 38 (8), 2343. https://doi.org/10.1364/josab.428809.

Okamoto, K., 2006. Fundamentals of Optical Waveguides. Academic Press.

Otterstrom, N.T., Behunin, R.O., Kittlaus, E.A., Wang, Z., Rakich, P.T., 2018. A silicon Brillouin laser. Science 360 (6393), 1113–1116. https://doi.org/10.1126/science.aar6113. https://www.science.org/doi/abs/10.1126/science.aar6113.

Pant, R., Poulton, C.G., Choi, D.Y., Mcfarlane, H., Hile, S., Li, E., Thevenaz, L., Luther-Davies, B., Madden, S.J., Eggleton, B.J., 2011. On-chip stimulated Brillouin scattering. Opt. Express 19 (9), 388–392. http://www.opticsinfobase.org/abstract.cfm?uri=oe-19-9-8285.

Piotrowski, J., Schmidt, M.K., Stiller, B., Poulton, C.G., Steel, M.J., 2021. Picosecond acoustic dynamics in stimulated Brillouin scattering. Opt. Lett. 46 (12), 2972–2975. https://doi.org/10.1364/OL.424701. http://opg.optica.org/ol/abstract.cfm?URI=ol-46-12-2972.

Primo, A.G., Carvalho, N.C., Kersul, C.M., Frateschi, N.C., Wiederhecker, G.S., Alegre, T.P.M., 2020. Quasinormal-mode perturbation theory for dissipative and dispersive optomechanics. Phys. Rev. Lett. 125 (23), 233601. https://doi.org/10.1103/PhysRevLett.125.233601.

Quesada, N., Sipe, J.E., 2017. Why you should not use the electric field to quantize in nonlinear optics. Opt. Letters 42 (17), 3443–3446.

Quesada, N., Helt, L.G., Menotti, M., Liscidini, M., Sipe, J.E., 2021. Beyond photon pairs: nonlinear quantum photonics in the high-gain regime. arXiv:1811.10219.

Rakich, P.T., Reinke, C., Camacho, R., Davids, P., Wang, Z., 2012. Giant enhancement of stimulated Brillouin scattering in the subwavelength limit. Phys. Rev. X 2 (1), 011008. https://doi.org/10.1103/PhysRevX.2.011008.

Safavi-Naeini, A.H., Chan, J., Hill, J.T., Alegre, T.P.M., Krause, A., Painter, O., 2012. Observation of quantum motion of a nanomechanical resonator. Phys. Rev. Lett. 108 (3), 033602. https://doi.org/10.1103/PhysRevLett.108.033602.

Santos, F.G.S., Espinel, Y.A.V., Luiz, G.O., Benevides, R.S., Wiederhecker, G.S., Alegre, T.P.M., 2017. Hybrid confinement of optical and mechanical modes in a bullseye optomechanical resonator. Opt. Express 25 (2), 508–529. https://doi.org/10.1364/OE.25.000508. http://opg.optica.org/oe/abstract.cfm?URI=oe-25-2-508.

Shi, Y., Cerjan, A., Fan, S., 2017. Acousto-optic finite-difference frequency-domain algorithm for first-principles simulations of on-chip acousto-optic devices. In: Nonlinear Optics, Optica Publishing Group. Paper NM2A.2 http://opg.optica.org/abstract.cfm?URI=NLO-2017-NM2A.2.

Sipe, J.E., Steel, M.J., 2016. A Hamiltonian treatment of stimulated Brillouin scattering in nanoscale integrated waveguides. New J. Phys. 18, 045004. https://doi.org/10.1088/1367-2630/18/4/045004. 1–21.

Sipe, J.E., Bhat, N.A.R., Chak, P., Pereira, S., 2004. Effective field theory for the nonlinear optical properties of photonic crystals. Phys. Rev. E Stat. Nonlin. Soft Matter. Phys. 69 (1), 20. https://doi.org/10.1103/PhysRevE.69.016604.

Smith, M.J.A., De Sterke, C.M., Wolff, C., Lapine, M., Poulton, C.G., 2017. Enhanced acousto-optic properties in layered media. Phys. Rev. B 96 (6), 064114. https://doi.org/10.1103/PhysRevB.96.064114.

Sturmberg, B.C.P., Dossou, K.B., Smith, M.J.A., Morrison, B., Poulton, C.G., Steel, M.J., 2018. NumBAT: the integrated, open source numerical Brillouin analysis tool. arxiv, 1811.10219. http://arxiv.org/abs/1811.10219.

Tomes, M., Carmon, T., 2009. Photonic micro-electromechanical systems vibrating at *X*-band (11-GHz) rates. Phys. Rev. Lett. 102, 113601. https://doi.org/10.1103/PhysRevLett.102.113601.

Vassallo, C., 1991. Optical Waveguide Concepts. Elsevier.

Velchev, I., Neshev, D., Hogervorst, W., Ubachs, W., 1999. Pulse compression to the sub-phonon lifetime region by half-cycle gain in transient stimulated Brillouin scattering. IEEE J. Quantum Electron. 35, 1812–1816. https://doi.org/10.1109/3.806589.

Vogel, W., Welsch, D.-G., 2006. Quantum Optics. John Wiley & Sons.

Wiederhecker, G.S., Dainese, P., Mayer Alegre, T.P., 2019. Brillouin optomechanics in nanophotonic structures. APL Photonics 4 (7), 071101. https://doi.org/10.1063/1.5088169.

Winful, H.G., 2013. Chirped Brillouin dynamic gratings for storing and compressing light. Opt. Express 21 (8), 10039–10047. https://doi.org/10.1364/OE.21.010039. https://www.osapublishing.org/oe/abstract.cfm?uri=oe-21-8-10039.

Wolff, C., Steel, M.J., Poulton, C.G., 2014. Formal selection rules for Brillouin scattering in integrated waveguides and structured fibers. Opt. Expr. 1094-4087. 22 (26), 32489. https://doi.org/10.1364/OE.22.032489.

Wolff, C., Gutsche, P., Steel, M.J., Eggleton, B.J., Poulton, C.G., 2015a. Power limits and a figure of merit for stimulated Brillouin scattering in the presence of third and fifth order loss. Opt. Express 23 (20), 26628–26638. https://doi.org/10.1364/OE.23.026628. http://opg.optica.org/oe/abstract.cfm?URI=oe-23-20-26628.

Wolff, C., Steel, M.J., Eggleton, B.J., Poulton, C.G., 2015b. Stimulated Brillouin scattering in integrated photonic waveguides: forces, scattering mechanisms and coupled mode analysis. Phys. Rev. A 92, 013836, 1–13. https://doi.org/10.1364/OE.22.029270.

Wolff, C., Stiller, B., Eggleton, B.J., Steel, M.J., Poulton, C.G., 2017. Cascaded forward Brillouin scattering to all stokes orders. New J. Phys. 19 (2), 023021. https://doi.org/10.1088/1367-2630/aa599e.

Wolff, C., Smith, M.J.A., Stiller, B., Poulton, C.G., 2021. Brillouin scattering—theory and experiment: tutorial. J. Opt. Soc. Am. B 38 (4), 1243–1269. https://doi.org/10.1364/JOSAB.416747. http://opg.optica.org/josab/abstract.cfm?URI=josab-38-4-1243.

Yariv, A., Yariv, P.E.E.A., Yeh, P., 1984. Optical Waves in Crystals: Propagation and Control of Laser Radiation. A Wiley Interscience Publication, Wiley, ISBN: 9780471091424. https://books.google.com.br/books?id=jjzxAAAAMAAJ.

Yuan, H., Wang, Y., Zhu, C., Zheng, Z., Lu, Z., 2019. Investigation of sub-phonon lifetime pulse amplification in active frequency matching stimulated Brillouin scattering. Opt. Express 27, 16661–16670. https://doi.org/10.1364/oe.27.016661.

CHAPTER THREE

The convergence of cavity optomechanics and Brillouin scattering

Mikołaj K. Schmidt[a,*], Christopher G. Baker[b], and Raphaël Van Laer[c]
[a]Macquarie University Research Centre in Quantum Engineering (MQCQE), MQ Photonics Research Centre, School of Mathematical and Physical Sciences, Macquarie University, Sydney, NSW, Australia
[b]Australian Research Council Centre of Excellence for Engineered Quantum Systems, School of Mathematics and Physics, The University of Queensland, St Lucia, QLD, Australia
[c]Department of Microtechnology and Nanoscience (MC2), Chalmers University of Technology, Gothenburg, Sweden
*Corresponding author: e-mail address: mikolaj.schmidt@mq.edu.au

Contents

1. Introduction

Until recently, there were two separate communities studying interactions between light and sound: those of *cavity optomechanics* and *Brillouin scattering*. These research fields developed separately for historical reasons.

Semiconductors and Semimetals, Volume 109
ISSN 0080-8784
https://doi.org/10.1016/bs.semsem.2022.04.005

Braginsky pioneered the study of the canonical cavity optomechanical structure in the 1960s: a Fabry–Pérot cavity with a movable mirror (shown in Fig. 1A), which forms a mechanical oscillator coupled to the light field. His work was geared toward pushing the limits of gravitational wave detectors (Braginsky and Manukin, 1967; Braginsky et al., 2001). Much earlier, in the 1920s, Brillouin and Mandelstam were concerned with studying interactions between light and thermal fluctuations in transparent bodies (Brillouin, 1922; Mandelstam, 1926).[a] Their work laid the basis for studying interactions between laser light and propagating GHz acoustic vibrations in optical fibers and, eventually, integrated waveguides (Eggleton et al., 2013). In a canonical Brillouin interaction platform, shown schematically in Fig. 1B, traveling light and GHz sound waves interact with one another in an optical fiber (Kobyakov et al., 2010)—typically in a backward scattering configuration where the energy and momentum difference between two optical waves roughly matches the energy and momentum of a GHz acoustic wave.

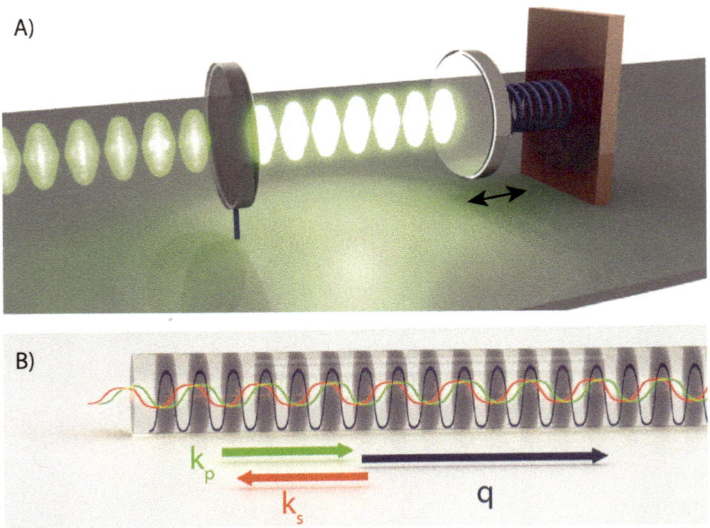

Fig. 1 Canonical cavity optomechanics and Brillouin scattering systems: (A) Fabry–Pérot cavity with fixed (*left*) and movable (*right*) end mirrors, where the optomechanical interaction is dominated by moving boundary/radiation pressure coupling; (B) Brillouin interaction between pump (*green*, momentum k_p) and Stokes (*red*, momentum k_S) photons and a traveling density wave in a fiber (*blue*) with momentum $q = 2\pi/\lambda$, mediated by photoelastic/electrostrictive coupling.

[a] For a detailed historical perspective, see chapter "Historical perspective and basic principles" by Poulton et al. of this book.

Why have these research fields converged recently? To answer this, we note that light-sound interactions may be classified according to the answers to a few questions:

1. In how many dimensions of space are the light and sound localized? Does the system have few or many electromagnetic and acoustic modes?
2. Do the mechanical waves have high or low group velocity?
3. What is the physical mechanism mediating the interaction?
4. What are the relative spatial and temporal decay rates of the light and sound?

Depending on the answers to these questions, systems were traditionally seen as either optomechanical cavities or as Brillouin scattering systems. While this distinction was usually not made explicitly in practice, up to roughly the year 2011, essentially all systems could be clearly classified according to these categories (Fig. 2). On the one hand, cavity optomechanics used to deal with three-dimensional confinement of electromagnetic and mechanical waves in few-moded devices (Fig. 2*i*), mechanical waves with negligible group velocity (Fig. 2*ii*), interactions mediated mostly by moving boundaries changing the optical properties (Fig. 2*iii*), and optical decay rates κ far exceeding the mechanical decay rates Γ (Fig. 2*iv*). On the other hand, the Brillouin field used to deal mostly with at most two-dimensional confinement in many-moded

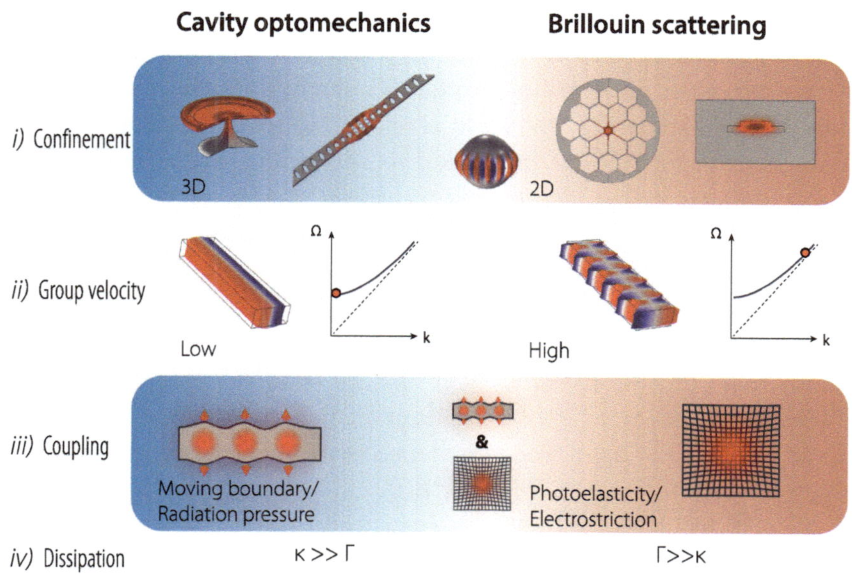

Fig. 2 Historical differences between cavity optomechanics and Brillouin scattering. These differences have blurred in recent years.

devices (Fig. 2*i*), mechanical waves with high group velocity (Fig. 2*ii*), interactions mediated mostly by strain locally changing the optical permittivity (Fig. 2*iii*), and mechanical decay rates far exceeding the optical decay rates (Fig. 2*iv*).

These distinctions between the fields have blurred, and started disappearing in the last decade, as experiments and theory have both shown that the categories defined by the above questions are mostly artificial, and that essentially identical core principles underpin both research fields. Indeed, both Brillouin scattering and cavity optomechanics describe three-wave-mixing effects in which two photons of slightly different frequencies interact with a mediating acoustic phonon in a dynamical feedback loop. Consequently, these fundamental similarities are reflected in the adopted mathematical formalisms and the observed physical effects. Unfortunately, a number of mostly superficial differences in the manifestation of the interactions have created different terminologies, creating a language barrier between the communities.

It is thus instructive to look back to the development of cavity optomechanics and Brillouin scattering over the last two decades, and highlight how the differences in physical platforms led to this historical divergence of terminology and formalisms. We provide this perspective in Section 1.1, and review the convergence of the systems explored in the two fields, highlighting how their modern realizations blur the division, in Section 2. In Section 3, we briefly recall the formulation used by the two communities presented in chapter "Theoretical formalisms for stimulated Brillouin scattering" by Wolff et al., and introduce formal mappings between these formalisms. In Section 4, we review several fundamental effects explored independently, and in parallel, by the two communities, including the spontaneous, and coherent opto-acoustic coupling effects.

We note that here we focus only on systems exploring the coupled guided light and acoustic phonons colocalized in opto-acoustic cavities or waveguides. We direct the reader to the following review articles: Safavi-Naeini et al. (2019), Aspelmeyer et al. (2014a), Eggleton et al. (2019), and Wiederhecker et al. (2019) for a more detailed and extensive perspective on the landscape of opto- and electromechanics.

1.1 Historical overview

Cavity optomechanics encompasses any form of interaction between electromagnetic waves—both at optical and microwave frequencies—and mechanical systems in structures where all the excitations are strongly confined in three dimensions.

A driving force behind the growth of the field was the development of large-scale interferometric gravitational wave detectors where light interacts with suspended masses in a fashion similar to Fig. 1A (Braginsky et al., 2001). The radiation pressure interaction between the confined light and the movable mirror leads to the well-known optomechanical interaction Hamiltonian (Aspelmeyer et al., 2014a):

$$\mathcal{H}_{\text{int}} = \hbar g_0 a^\dagger a \left(b + b^\dagger\right), \tag{1}$$

which describes coupling between quantized optical and acoustic fields, described through the ladder operators a and b. As we discuss in detail in Section 4, this interaction is responsible for phenomena such as optical bistability (Dorsel et al., 1983) and the optical spring (Sheard et al., 2004), as well as dynamical backaction effects, including mechanical amplification and cooling (Arcizet et al., 2006).

In 2005 (Carmon et al., 2005; Kippenberg et al., 2005), these backaction effects were observed in smaller and higher-frequency systems: ultra-high-Q optical microcavities called microtoroids. These are silica microdisks (radius $R \simeq 10^{-5}$ m), the outer circumference of which is reflown using a CO_2 laser, removing fabrication-induced sidewall roughness responsible for scattering losses (Armani et al., 2003). This process results in a nearly atomically smooth annular outer boundary, allowing the trapping of photons for extended periods of time by total internal reflection in optical whispering gallery modes (WGMs) with quality factors in excess of 10^8. These experiments focused mainly on the coupling of (traveling or standing) optical WGMs to the acoustic radial breathing mode (RBM) of the resonator, where the resonator expands and contracts radially in and out with a frequency in the tens of MHz. The focus on the RBM is natural, as all WGMs of the structure are naturally well-coupled to its motion, described by a generalized spatial coordinate x, with a coupling rate[b] $G = \partial\omega_0/\partial x \simeq \omega_0/R$ (Ding et al., 2010). In contrast, while backward and forward scattering interactions with traveling acoustic waves can be achieved with WGM resonators, these require careful control of the device size to match its free spectral range (FSR) at least roughly to the acoustic frequency (see, e.g., Lee et al. (2012)) or careful selection of a pair of WGMs whose frequencies and azimuthal number satisfy energy and momentum matching criteria (see, e.g., the optical cooling of a 100 MHz mechanical mode in a silica microsphere via photoelastic coupling

[b] Throughout this chapter, we will differentiate the optomechanical coupling G (with $[G] =$ rad·m^{-1}·s^{-1}) from Brillouin gain G_B (with $[G_B] = (\text{W}\cdot\text{m})^{-1}$) by subscript B.

by Bahl et al., 2012a). These conditions are both difficult to achieve with microtoroids, which combine sparse optical spectra and poorly controlled physical dimensions due to shrinkage in the reflow process.

These and similar systems then became the workhorses of modern on-chip cavity optomechanics, their small size, and high optical Q enabling them to operate in a regime where the mechanical frequency becomes comparable (and even larger) than the optical decay rate, or linewidth, the so-called *resolved sideband regime* leading to efficient dynamical backaction. This led to a successful push toward coherent and quantum regimes, realizing several feedback effects such as *resolved sideband cooling* and *optomechanically induced transparency* (OMIT) (Weis et al., 2010), both of which we touch upon in this chapter. We note that while the RBM's radial displacement u_r is maximal near the perimeter, its strain—defined by the spatial derivative of u_r—is largest near the center of the disk, and near zero at the circumference where the WGM is located. As a consequence, coupling to RBMs in microtoroids and microdisks—as well as to mirror motion in Fabry–Pérot cavities, and a host of other optomechanical implementations such as "membrane-in-the-middle" schemes (Thompson et al., 2008) and cavity electromechanics with microwave photons (Teufel et al., 2011)—is dominated by *moving boundary/radiation pressure coupling*. For this reason, in its early days much of the field focused solely on the coupling of light to lower-frequency (<GHz) stationary acoustic waves via moving-boundary coupling.

Only later, by designing even smaller cavities with $\sim \lambda^3$ mode volumes made from high-refractive-index materials such as silicon and gallium arsenide were regions of high optical field colocated with regions of high mechanical *strain*. This was the case for optomechanical crystals, made of silicon (Chan et al., 2011, 2012; Safavi-Naeini et al., 2011), gallium arsenide (Balram et al., 2014), or diamond (Kipfstuhl et al., 2014), which mediated highly efficient interactions between near-infrared light and GHz mechanical motion via *both* moving boundaries and photoelasticity. Similarly, wavelength-scale GaAs WGM resonators (Ding et al., 2011) with mechanical frequencies in excess of 1 GHz operate in a regime in which the photoelastic coupling matches and even exceeds the moving boundary coupling (Baker et al., 2014).

Brillouin scattering, on the other hand, relied until recently mostly on the dependence of the optical refractive index on mechanical strain (i.e., the *photoelastic effect*) to mediate interactions between *traveling* optical and acoustic waves (as illustrated in Fig. 1B, where the well-confined optical modes are insensitive to boundary motion where the field is minimal).

In addition, it largely dealt with acoustic frequencies in the GHz to tens of GHz range, as dictated by the simultaneous frequency- and momentum-matching conditions. Coupling based on moving boundaries changing the effective optical refractive index was predicted and experimentally demonstrated in Brillouin scattering in the past decade, as optomechanical waveguides with submicron cross sections were realized (Rakich et al., 2012; Shin et al., 2013; Van Laer et al., 2015a, b). In these devices, competition between the two interaction mechanisms, i.e., the moving-boundary and photoelastic effect, must be properly engineered. In another example, Van Laer et al. (2015b) observed both interactions between copropagating and counterpropagating optical modes in a single silicon nanowire via slow and fast GHz mechanical modes, respectively. Some of these GHz waveguide modes turned out to be close analogs to the mechanical modes found in GHz optomechanical crystals (Chan et al., 2012; Van Laer, 2016; Van Laer et al., 2016). Furthermore, the boundary between traveling and stationary modes was blurred by embracing cryogenic systems with phonon decay lengths matching, and even exceeding the lengths of the resonant cavities, enabling mechanical feedback.

The convergence of the two fields is well illustrated by the following hybrid systems which combine the traditional optomechanical and Brillouin-type coupling interactions within the same microstructure. Han et al. (2013) demonstrated a silica hollow whispering-gallery microresonator, which functioned as multimode optomechanical pressure sensor, with acoustic modes at 16 MHz and 11 GHz, simultaneously driven by moving boundary and photoelastic coupling, respectively. More recently, He et al. (2020) showed active switching between standing and traveling wave interactions within the same device, by controlling the standing or traveling-wave nature of the optical WGMs and sound waves in a superfluid microresonator. In the latter regime, the authors observed phonon lasing and, consequently, a strong coupling between the counterpropagating optical WGMs mediated by a large coherent phonon amplitude.

2. Converging fields: Basic physics and nomenclature

2.1 Three-wave mixing in cavities and waveguides

We continue our journey by briefly illustrating the basic three-wave-mixing nature of the interaction in few-moded cavities and in many-moded waveguides. Beyond the more formal treatment provided below, we note that the nature of the three-wave mixing can also be intuitively understood in terms of a mass-on-a-spring model (Kippenberg and Vahala, 2007; Van Laer et al., 2016). The mass has a certain resonance frequency and its motion is damped by a viscous force. Suppose we measure the position of the freely moving mass. Next, we feed back this information by driving the mass with a force that is in phase with the velocity. This results in an effectively lower decay rate for the oscillator, since this force cannot be distinguished from the intrinsic viscous damping. On the other hand, a feedback force in antiphase with the velocity yields an effectively larger decay rate. Similarly, a force in phase with the position generates a shift of the spring constant. Variations on this theme are responsible for all three-wave-mixing effects involving light and sound. This deceptively simple feedback loop generates a wealth of interesting effects.

This feedback loop can be seen in the interaction Hamiltonians of both waveguides and cavities. As shown in Section 3.1 and chapter "Theoretical formalisms for stimulated Brillouin scattering" by Wolff et al., the interaction Hamiltonian in a few-moded cavity—in this case supporting two optical modes 1 and 2—is given by

$$\mathcal{H}^{\text{cavity}}_{\text{interaction}} = \hbar g_0 \left(a_1^\dagger a_2 b + \text{h.c.} \right), \tag{2}$$

where we show only the essential three-wave-mixing terms containing the optical annihilation operators a_1 and a_2 and a mechanical mode with annihilation operator b. This Hamiltonian shows the essential nature of the interaction: two electromagnetic modes and one mechanical mode are exchanging quanta through creation and annihilation processes. For example, a photon in mode a_1 can be created, while a photon in mode a_2 and a phonon in b are annihilated. The mechanical motion described by operator b is being measured by the optical modes a_1 and a_2, and this information is being written back onto the mechanical oscillator with some delay through a beat note in the optical forces described by $a_1^\dagger a_2$ acting back on the mechanical oscillator. This feedback process has an interaction strength g_0, which is set by the efficiency of both the readout and the feedback.

A cavity can be constructed from a waveguide by looping the waveguide back on itself or terminating the waveguide with mirrors. Therefore, the Hamiltonian of an optomechanical cavity given in Eq. (2) and physics governing its dynamics are closely related to the Hamiltonian and physics of many-moded waveguides (Safavi-Naeini et al., 2019; Van Laer et al., 2016). Indeed, it is shown in chapter "Theoretical formalisms for stimulated Brillouin scattering" by Wolff et al. that the waveguide Hamiltonian can be expressed as (Sipe and Steel, 2016)

$$\mathcal{H}^{\text{waveguide}}_{\text{interaction}} = \frac{\hbar g_{0|\beta}}{\sqrt{2\pi}} \int\int \mathrm{d}k\, \mathrm{d}\beta\, a^{\dagger}_{1,\beta+k} a_{2,k} b_{\beta} + \text{h.c.} \tag{3a}$$

$$= \hbar g_{0|\beta} \int \mathrm{d}z\, a_1^{\dagger}(z) a_2(z) b(z) + \text{h.c.}, \tag{3b}$$

where we give the momentum-space form in the first line and real-space form in the second line. In this case, the annihilation operators refer to either modes with specific wavevectors (β, k, and $k+\beta$), or locations (z) in space along an extended waveguide. Similar to the cavity Hamiltonian (Eq. 2), this Hamiltonian describes basic three-wave-mixing interactions which can be understood in our mass-on-a-spring model. This also implies that the waveguide's interaction rate $g_{0|\beta}$ can be related to the cavity's interaction rate g_0 (Van Laer et al., 2016). This correspondence is explored further in Section 3.3. There we show that Eq. (2) can be derived from Eq. (3a) in the high-finesse limit via a so-called *mean-field transition.* The connection between waveguides and cavities was explored in depth in Van Laer et al. (2016) as well as in Rakich and Marquardt (2018) and briefly in Safavi-Naeini and Painter (2010).

This illustrates briefly what can be shown more rigorously (Van Laer et al., 2016), i.e., that all the dynamical effects taking place in cavities must have closely related counterparts in extended waveguides. That is, few-moded cavity effects like optomechanically induced transparency and sideband cooling have equivalents in a many-moded waveguide. Conversely, many physical effects taking place in a waveguide, such as Brillouin slow light and optical amplification, find counterparts in high-finesse cavities.

The waveguide Hamiltonian (Eq. 3a) has a more complex form than the cavity Hamiltonian (Eq. 2) as it sums over interactions between many possible mode pairs in momentum space. It captures interactions taking place in a continuum in the limit of a translationally invariant waveguide. This introduces a limitation to the equivalence between cavity- (Eq. 2) and (Eq. 3) light-sound interactions. Certain many-moded physical effects possible in an extended waveguide do not have a clear equivalent in a few-moded cavity. For instance, counter-propagating optical pulses could briefly interact with a traveling acoustic pulse as they pass each other at a certain point along an extended waveguide (Zhu et al., 2007). Such spatial information is lost in a high-finesse cavity. So while every physical effect seen in few-moded and high-finesse cavities has a counterpart in a waveguide, the converse is not necessarily true. This is due to the increased number of modes which selectively interact as determined by phase-matching selection rules.

2.2 Naming conventions

Before continuing on to a more rigorous treatment, we aim to clarify a few naming conventions. Over time some terms have become loaded with additional layers of meaning which are unknown to those outside the community. We aim to reduce this language barrier.

Interaction mechanisms: Forces and scattering. The naming of "radiation pressure" and "electrostriction" for the optical forces can be confusing, as electrostrictive stresses are also a form of pressure due to radiation. While both refer to pressure exerted by electromagnetic radiation, the "radiation pressure" force describes what was historically termed the *pressure of light*, and was first experimentally measured by reflecting light off a mirror affixed to a torsion balance (Lebedew, 1901; Nichols and Hull, 1901). On the scattering side, a confined photon is sensitive to changes to its boundary conditions, as well as to changes in the medium's permittivity, caused by bulk strain and the material's photoelastic coefficients. These two coupling mechanisms are referred to as *moving boundary* and *photoelastic* couplings. To avoid any confusion, we will refer to the optical scattering side of the interaction as much as possible, and therefore to *moving boundary* (MB) and *photoelastic* (PE) couplings, rather than to their respective backaction forces *radiation pressure* (RP) and *electrostriction* (ES). This has the additional advantage of clarifying ambiguities related to the Abraham–Minkowski debate on the momentum of a photon in a medium (Nelson, 1991; Wolff et al., 2015).

The cavity optomechanics community often uses the term "radiation pressure" to mean *any* force exerted by light (Aspelmeyer et al., 2014b). In the Brillouin community, however, this term is typically reserved for the moving-boundary coupling. Here, we mostly follow the latter convention.

Brillouin, optomechanics, waveguides, and cavities. We intend to no longer use "Brillouin" or "optomechanics" to refer to specific properties or parameter ranges of the light-sound interactions in a certain system as defined by the four questions at the beginning of Section 1. Instead, all physical effects resulting from light-sound interactions, whether in waveguides or cavities, may be called Brillouin and optomechanical effects. Terms involving "light-sound," "opto-acoustic," "photon-phonon," "waveguide," and "cavity" do not suffer from historical ambiguity and are often preferred for clarity.

Pairing and naming of physical effects. Over the years a host of duplicate names have been given to identical or closely related physical effects. For example, the following effects can be paired up:

1. *Optomechanically induced transparency (OMIT) ↔ Brillouin-scattering induced transparency (BSIT).* These are both forms of electromagnetically induced transparency (EIT): electromagnetic fields create transparency windows

via light-matter interaction (Safavi-Naeini et al., 2011; Weis et al., 2010) (see Section 4.2 for detailed discussion),

2. *Reversed dissipation hierarchy of cavity optomechanics* ↔ *Regular dissipation hierarchy in waveguides*. When the temporal acoustic decay rate exceeds the temporal optical decay rate, this has been called the "reversed" dissipation regime of cavity optomechanics (Nunnenkamp et al., 2014). However, this is the canonical situation for the spatial decay in waveguides (Van Laer et al., 2016) (see Section 3.4.1),
3. *Brillouin laser* ↔ *Optical laser*. The term Brillouin laser has traditionally referred to an optical laser based on Brillouin scattering. However, it may mean acoustic laser if "Brillouin" is taken in a broader sense (see Section 4.3),
4. *Brillouin feedback loop* ↔ *Optomechanical backaction*. These both refer to the intrinsic three-wave-mixing feedback loop, where forces and scattering can reinforce each other,
5. *Optomechanical wavelength conversion* ↔ *Brillouin cross-phase modulation*. An intensity-modulated optical wave at one wavelength can phase-modulate an optical signal at a far different wavelength. This has been called cross-phase modulation in the Brillouin community (Van Laer et al., 2015b), and optomechanical wavelength conversion in the cavity optomechanics community (Hill et al., 2012). They both stem from an effective four-wave-mixing interaction resulting from eliminating an acoustic intermediary (Safavi-Naeini et al., 2019).

We will discuss many of these cases in detail in the following sections. They all result from the fundamental three-wave-mixing interaction. It is often the relative size of the coupling strength as well as optical and acoustic dissipation that determines which of these effects is observed in a certain situation.

3. Theoretical description of coupling between light and sound in waveguides and cavities

There are a number of interaction mechanisms enabling light-sound coupling both in waveguides and cavities. This includes moving-boundary (MB), photoelastic (PE), roto-optic (RO), and moving polarization (MP) effects. These physical coupling mechanisms are discussed in detail in chapter "Theoretical formalisms for stimulated Brillouin scattering" by Wolff et al. This convergence is reflected in the mathematical formalisms which can be used to characterize both systems. In particular, by embracing the Hamiltonian description of the interacting optical and acoustic modes (summarized for cavity optomechanics in Aspelmeyer et al., 2014b, and for Brillouin scattering in Sipe and Steel, 2016), we can readily perform the canonical quantization of the system, and trace, for example, how single photons can scatter to generate pairs of localized or propagating photons and phonons. From there, we can study the dynamics of the quantum states of optical and acoustic modes, by using the Heisenberg equations, and map systems to the classical problem of stationary, or propagating intertwined optical and acoustic waves.

In Section 3.1, we present a detailed discussion of the formalism of cavity optomechanics, revisiting concepts presented in chapter "Theoretical formalisms for stimulated Brillouin scattering" by Wolff et al. In Section 3.2, we apply the same formalism to identify the Hamiltonian and Heisenberg equations governing the opto-acoustic interaction in waveguides. Finally, in Section 3.3 we review proposals for mapping these formulations, and use them to connect the fundamental figures of merit used in both communities in Section 3.4.

3.1 Cavity optomechanics

We begin the derivation of the optomechanical cavity Hamiltonian by decomposing the electric and acoustic displacement fields $\boldsymbol{E}$ and $\boldsymbol{U}$ into eigenmodes of the optical ($\boldsymbol{e}^{(n)}(\boldsymbol{r})$) and acoustic ($\boldsymbol{u}^{(m)}(\boldsymbol{r})$) wave equations, respectively,

$$\boldsymbol{E}(\boldsymbol{r},t) = \sum_n a_n(t)\exp(-\mathrm{i}\omega_n t)\boldsymbol{e}^{(n)}(\boldsymbol{r}) + \text{h.c.}, \tag{4a}$$

$$\boldsymbol{U}(\boldsymbol{r},t) = \sum_m b_m(t)\exp(-\mathrm{i}\Omega_m t)\boldsymbol{u}^{(m)}(\boldsymbol{r}) + \text{h.c.} \tag{4b}$$

The canonical quantization promotes the classical envelopes a_n and b_m to quantum-mechanical operators, obeying the commutation relations

$$\left[a_n(t), a_{n'}^\dagger(t)\right] = \delta_{nn'}, \quad \left[b_m(t), b_{m'}^\dagger(t)\right] = \delta_{mm'}, \tag{5a}$$

and normalizing the eigenmode profiles $\boldsymbol{e}^{(n)}$ and $\boldsymbol{u}^{(m)}$ as in

$$\int \mathrm{d}^3\boldsymbol{r}\, \varepsilon_0 \varepsilon(\boldsymbol{r}) |\boldsymbol{e}^{(n)}(\boldsymbol{r})|^2 = \frac{\hbar\omega_n}{2}, \tag{6a}$$

$$\Omega_m^2 \int \mathrm{d}^3\boldsymbol{r}\, \rho(\boldsymbol{r}) |\boldsymbol{u}^{(m)}(\boldsymbol{r})|^2 = \frac{\hbar\Omega_m}{2}. \tag{6b}$$

This formulation is consistent with that adopted in the section of chapter "Theoretical formalisms for stimulated Brillouin scattering" by Wolff et al.

describing optomechanics in cavities, and has a particularly appealing property of retaining the physical units of the operators corresponding to the electric fields ($[\boldsymbol{E}] = [\boldsymbol{e}^{(n)}] = \mathrm{V} \cdot \mathrm{m}^{-1}$) and acoustic displacement fields ($[\boldsymbol{U}] = [\boldsymbol{u}^{(m)}] = \mathrm{m}$). Normalization constants introduced in Eqs. (14) and 20) in chapter "Theoretical formalisms for stimulated Brillouin scattering" by Wolff et al. simplify to $\mathcal{E}_{b,m} = \hbar\Omega_m$ and $\mathcal{E}_n = \hbar\omega_n$.

We can write down the Hamiltonian of the noninteracting optical and acoustic system as

$$\mathcal{H}_0^{(\mathrm{OM,cav})}/\hbar = \sum_n \omega_n a_n^\dagger a_n + \sum_m \Omega_m b_m^\dagger b_m. \tag{7a}$$

3.1.1 Effective quantities

Before we continue to describe the interaction of the optical and acoustic modes, let us define two quantities often used to approximately characterize the *spatial confinement* and *inertia* of the modes: the effective optical mode volume $V_{\mathrm{eff}}^{(n)}$ and effective acoustic mode mass $m_{\mathrm{eff}}^{(m)}$ introduced by rewriting the left-hand sides of Eq. (6) as

$$\int \mathrm{d}\boldsymbol{r}\, \varepsilon_0 \varepsilon(\boldsymbol{r}) |\boldsymbol{e}^{(n)}(\boldsymbol{r})|^2 = V_{\mathrm{eff}}^{(n)} \max_{\boldsymbol{r}} \left[\varepsilon_0 \varepsilon(\boldsymbol{r}) |\boldsymbol{e}^{(n)}(\boldsymbol{r})|^2\right], \tag{8a}$$

$$\Omega_m^2 \int \mathrm{d}\boldsymbol{r}\, \rho(\boldsymbol{r}) |\boldsymbol{u}^{(m)}(\boldsymbol{r})|^2 = \Omega_m^2 m_{\mathrm{eff}}^{(m)} \max_{\boldsymbol{r}} \left[|\boldsymbol{u}^{(m)}(\boldsymbol{r})|^2\right]. \tag{8b}$$

Furthermore, for the acoustic mode we also define the *zero-point fluctuations amplitude*

$$x_{\mathrm{zpf}}^{(m)} = \sqrt{\frac{\hbar}{2 m_{\mathrm{eff}}^{(m)} \Omega_m}}, \tag{9}$$

which—assuming normalization in Eq. (8b), is equal, and thus often used interchangeably, with $\max_{\boldsymbol{r}}\left[|\boldsymbol{u}^{(m)}(\boldsymbol{r})|\right]$. For other normalization schemes, this relationship might be more complex.

The above equations should be treated as definitions of the effective quantities, and thus yield identical numerical values, and the dimensions ($[V_{\mathrm{eff}}^{(n)}] = \mathrm{m}^3$, $[m_{\mathrm{eff}}^{(m)}] = \mathrm{kg}$), for any normalization scheme.

3.1.2 Optomechanical interaction Hamiltonian

In a canonical optomechanical system, represented in Fig. 1A, the frequency of the optical cavity mode shifts with the displacement of one of its mirrors. In the lowest-order approximation, this shift is proportional to the displacement, and so the optomechanical coupling between optical mode a and an acoustic mode b is defined as an optical frequency shift divided by the mirror displacement, conventionally denoted as G. In the quantum-mechanical formulation of this problem, the displacement of the mirror is promoted to an operator $x = x_{zpf}(b^\dagger + b)$, prompting us to introduce a *per phonon* optomechanical coupling $g_0 = Gx_{zpf}$.

The optomechanical coupling mechanisms, and the corresponding coefficients, are carefully derived and discussed in several recent reviews on cavity optomechanics (Safavi-Naeini et al., 2019; Wiederhecker et al., 2019). Here, we choose to only highlight the expression for the photo-elastic/electrostrictive coupling in two-mode optomechanical coupling, between an optical mode with profile $\boldsymbol{e}^{(n)}$ and an acoustic mode with profile $\boldsymbol{u}^{(m)}$, to match the corresponding expression describing coupling in Brillouin systems, given by Eq. (50b) in chapter "Theoretical formalisms for stimulated Brillouin scattering" by Wolff et al. Since there is no ambiguity about the modes, we drop the superscripts (n/m). We also write it first adopting the expression from Wiederhecker et al. (2019), which does not implicitly rely on the properties of a particular normalization scheme:

$$g_0^{(PE)} = -\frac{\omega}{2}\frac{x_{zpf}}{\max_{\boldsymbol{r}}[|\boldsymbol{u}(\boldsymbol{r})|]}\frac{\int \mathrm{d}^3\boldsymbol{r}\ [\boldsymbol{e}(\boldsymbol{r})]^* \cdot \overline{\delta\varepsilon}(\boldsymbol{r}) \cdot \boldsymbol{e}(\boldsymbol{r})}{\int \mathrm{d}\boldsymbol{r}\ \varepsilon_0\varepsilon(\boldsymbol{r})|\boldsymbol{e}(\boldsymbol{r})|^2}, \tag{10}$$

with units of $[g_0^{(PE)}] = \mathrm{rad}\cdot\mathrm{s}^{-1}$. The local change of the relative permittivity $\overline{\delta\varepsilon} = -\varepsilon_0 n^4 \overleftrightarrow{p} : s$ is defined by the local refractive index n, photoelastic tensor $\overleftrightarrow{p}$, and the *normalized* strain s of the acoustic mode. Adopting the particular normalization described by (Eq. 6), we can simplify that expression to read[c]

$$g_0^{(PE)} = \frac{1}{\hbar\varepsilon_0}\int \mathrm{d}^3\boldsymbol{r}\ [\boldsymbol{d}(\boldsymbol{r})]_i^*[\boldsymbol{d}(\boldsymbol{r})]_j p_{ijkl}(\boldsymbol{r}) s_{kl}(\boldsymbol{r}). \tag{12}$$

with $\boldsymbol{d}$ the normalized electric displacement field. Adding the contributions from the other physical effects governing the interaction:

[c] For the different convention of expanding the fields into eigenmodes, similar to that given in Eq. (29) for waveguides, we have $x_{zpf} = \sqrt{\hbar\Omega/2}\max[|\boldsymbol{u}(\boldsymbol{r})|]$, and

$$g_0^{(PE)} = \frac{\omega}{2}\sqrt{\frac{\hbar\Omega}{2}}\int \mathrm{d}^3\boldsymbol{r}\ [\boldsymbol{d}(\boldsymbol{r})]_i^*[\boldsymbol{d}(\boldsymbol{r})]_j p_{ijkl}(\boldsymbol{r}) s_{kl}(\boldsymbol{r}). \tag{11}$$

$$g_0 = g_0^{(\mathrm{PE})} + g_0^{(\mathrm{MB})} + g_0^{(\mathrm{RO})} + g_0^{(\mathrm{MP})}, \tag{13}$$

we can write down the Hamiltonian of a two-mode optomechanical system:

$$\mathcal{H}^{(\mathrm{OM,cav})}/\hbar = \mathcal{H}_0^{(\mathrm{OM,cav})}/\hbar + g_0 a^\dagger a\left(b + b^\dagger\right). \tag{14}$$

One can also identify a straightforward extension of this formulation to a three-mode picture, where two optical modes (n and n') interact with an acoustic mode. Using the mode profiles normalized as above, the PE coupling coefficient is

$$g_{0,nn'}^{(\mathrm{PE})} = \frac{1}{\hbar\varepsilon_0}\int \mathrm{d}^3\boldsymbol{r}\, \left[\boldsymbol{d}^{(n)}(\boldsymbol{r})\right]_i^* \left[\boldsymbol{d}^{(n')}(\boldsymbol{r})\right]_j p_{ijkl}(\boldsymbol{r}) s_{kl}(\boldsymbol{r}), \tag{15}$$

yielding a Hamiltonian that includes both self- and cross-mode couplings

$$\begin{aligned} \mathcal{H}^{(\mathrm{OM,cav})}/\hbar = \mathcal{H}_0^{(\mathrm{OM,cav})}/\hbar + \sum_n g_{0,nn} a_n^\dagger a_n\left(b + b^\dagger\right) \\ + \sum_{n \neq n} g_{0,nn'}\left(a_n^\dagger a_{n'} + a_n a_{n'}^\dagger\right)\left(b + b^\dagger\right). \end{aligned} \tag{16a}$$

3.1.3 Heisenberg equations for two-mode cavity optomechanics

The dynamics of a cavity optomechanical system (Eq. 14) can be expressed by the Heisenberg equation for the time-dependent operators $a(t)$ and $b(t)$

$$\frac{\mathrm{d}}{\mathrm{d}t}a(t) = -\frac{i}{\hbar}\left[a(t), \mathcal{H}^{(\mathrm{OM,cav})}\right] = -i\omega a(t) + ig_0 a(t)\left[b(t) + b^\dagger(t)\right], \tag{17a}$$

$$\frac{\mathrm{d}}{\mathrm{d}t}b(t) = -\frac{i}{\hbar}\left[b(t), \mathcal{H}^{(\mathrm{OM,cav})}\right] = -i\Omega b(t) + ig_0 a^\dagger(t) a(t), \tag{17b}$$

thus, as expected, the optomechanical coupling induces a frequency shift of the optical resonance proportional to the acoustic mode displacement $x \propto b + b^\dagger$, and an optical force proportional to the photon number $n = a^\dagger a$, acting on the acoustic mode.

To complete this description, we need to account for the interaction with the environment: the coupling to the external optical control fields, to the thermal and vacuum fluctuations, and all other dissipation and associated noise channels that lead to the broadening of the spectrum of the optical and acoustic modes. To this end, for the optical mode, we separate the dissipation rates into contributions associated with input/output coupling, and internal losses: $\kappa = \kappa_{\mathrm{ex}} + \kappa_0$, and include into Eq. (17a) the following terms: $-a\kappa/2$ for total losses, $\sqrt{\kappa_0} f_{\mathrm{in}}$ for noise associated with internal losses, and $\sqrt{\kappa_{ex}}a_{in}$ for the control fields and external noise. In particular, the a_{in} optical input term would typically describe a coherent monochromatic drive due to laser source with input power P_{in} and frequency ω_L,

$$a_{in} = e^{-i\omega_L t}\sqrt{\frac{P_{in}}{\hbar\omega_L}} + d_{in}, \tag{18}$$

and noise contributions d_{in} which satisfy

$$\langle d_{in}(t)\rangle = 0, \quad \left\langle d_{in}(t)d_{in}^{\dagger}(t')\right\rangle = \delta(t - t'). \tag{19}$$

Using the input-output theory, the field transmitted, or reflected from the optomechanical cavity, is then described by $a_{out} = a_{in} - \sqrt{\kappa_{ex}}a$ (Aspelmeyer et al., 2014a).

To complete this description, we similarly extend Eq. (17b). Assuming no coherent driving of the mechanical mode, we characterize the external noise as

$$\langle b_{in}(t)\rangle = 0, \quad \left\langle b_{in}(t)b_{in}^{\dagger}(t')\right\rangle = \left(1 + n_b^{th}\right)\delta(t - t'). \tag{20}$$

Here, the mechanical mode is assumed to be coupled to a large, Markovian thermal reservoir with temperature T, and population given by the Planck distribution $n_b^{th} = \left[\exp\left(\hbar\Omega/k_B T\right) - 1\right]^{-1}$ which for $\hbar\Omega \ll k_B T$ (e.g., for GHz phonons at room temperature) is approximately $k_B T/\hbar\Omega$. The same reservoir yields negligible thermal contribution to the optical noise d_{in} since $\hbar\omega \gg k_B T$, and hence is dropped in Eq. (19).

Due to the nonlinear character of the above equations, one cannot identify a *finite* set of operators coupled via the Heisenberg–Langevin equations. This problem can be addressed by linearizing the optomechanical interaction around the coherent amplitudes of the optical and acoustic fields.

3.1.3.1 Linearizing the optomechanical interaction

In the simplest case of a continuous driving of the optical mode described above, the coherent amplitudes of the optical and acoustic modes can be found by moving to the frame rotating with laser frequency ω_L, considering the steady-state solution of the Heisenberg–Langevin equations, and calculating the expectation values of $\alpha = \langle a\rangle$[d] and $\beta = \langle b\rangle$:

[d] Through an unfortunate, but persistent convention, the Brillouin community uses α to denote both the coherent amplitudes of the optical fields ($\alpha_{p/s}$) and spatial decay rates ($\alpha_{p/s/b}$).

$$\alpha = \frac{\sqrt{\kappa_{\text{ex}}}\langle a_{\text{in}}\rangle}{\kappa/2 + i\Delta'}, \quad \beta = \frac{g_0|\alpha|^2}{\Omega - i\Gamma/2}, \tag{21}$$

where $\Delta' = \omega_a - \omega_L - 2g_0\text{Re}(\beta)$ is the effective detuning, which accounts for the detuning of the laser from an empty cavity $\Delta = \omega_a - \omega_L$,[e] and a static displacement of the cavity by $- x_{\text{zpf}}\text{Re}(\beta)$. These nonlinear equations can describe phenomena such as *static optomechanical instability* (Dorsel et al., 1983). They also allow us to formulate the Heisenberg–Langevin equations for the *fluctuations* $\delta a = a - \alpha$ and $\delta b = b - \beta$. Dropping the terms with quadratic dependence on fluctuations $\delta a^\dagger \delta b$, $\delta a^\dagger \delta b^\dagger$, and $\delta a^\dagger \delta a$, we find from Eq. (17)

$$\frac{\mathrm{d}}{\mathrm{d}t}\delta a(t) = -\left(\frac{\kappa}{2} + i\Delta'\right)\delta a(t) + ig\left[\delta b(t) + \delta b^\dagger(t)\right] + \sqrt{\kappa_{\text{ex}}}d_{\text{in}}(t) + \sqrt{\kappa_0}\,f_{\text{in}}, \tag{22a}$$

$$\frac{\mathrm{d}}{\mathrm{d}t}\delta b(t) = -\left(\frac{\Gamma}{2} + i\Omega\right)\delta b(t) + ig\left[\delta a(t) + \delta a^\dagger(t)\right] + \sqrt{\Gamma}b_{\text{in}}(t). \tag{22b}$$

The product $g = g_0|\alpha|$ is the *effective or pump-enhanced optomechanical coupling*, and the linearized equations of motion can be mapped to the linearized optomechanical Hamiltonian in a frame rotating with frequency ω_L:

$$\mathcal{H}^{(\text{OM,lin})} = \hbar\Delta'\delta a^\dagger\delta a + \hbar\Omega\delta b^\dagger\delta b - \hbar g(\delta a^\dagger + \delta a)(\delta b + \delta b^\dagger). \tag{23}$$

3.1.4 Three-mode interaction

Optomechanical systems can be also realized in three-mode system (a_1, a_2, b). Neglecting the dispersive terms in Eq. (16), and defining $g_0 = g_{0,12}$, we find

$$\frac{\mathrm{d}}{\mathrm{d}t}a_1(t) = -i\omega_1 a_1(t) + ig_0 a_2(t)\left[b(t) + b^\dagger(t)\right], \tag{24a}$$

$$\frac{\mathrm{d}}{\mathrm{d}t}a_2(t) = -i\omega_2 a_2(t) + ig_0 a_1(t)\left[b(t) + b^\dagger(t)\right], \tag{24b}$$

$$\frac{\mathrm{d}}{\mathrm{d}t}b(t) = -i\Omega b + ig_0\left[a_1^\dagger(t)a_2(t) + a_1(t)a_2^\dagger(t)\right]. \tag{24c}$$

In a typical scenario, we include into these equations noise terms as in Eqs. (19)–(20), and assume that the optical mode a_1 is driven by a monochromatic optical field, establishing a large coherent state with amplitude $\alpha_1 \exp(-i\omega_L t)$. In the absence of any other coherent driving fields, the

[e] In the cavity optomechanics literature, this detuning is often defined as $\Delta = \omega_L - \omega_a$.

two remaining modes a_2 and b do not develop a coherent amplitude (until they reach lasing thresholds, see Section 4.3) and, in the frame rotating with ω_L, Eqs. (24b) and (24c) can be linearized around α_1, and expressed as

$$\frac{\mathrm{d}}{\mathrm{d}t}a_2(t) = -\left(\frac{\kappa_2}{2} + i\omega_2 - i\omega_L\right)a_2(t) + ig\left[b(t) + b^\dagger(t)\right] + \sqrt{\kappa_{\mathrm{ex},2}}d_{\mathrm{in},2}(t) + \sqrt{\kappa_{0,2}}f_{\mathrm{in},2}, \tag{25a}$$

$$\frac{\mathrm{d}}{\mathrm{d}t}b(t) = -\left(\frac{\Gamma}{2} + i\Omega\right)b + ig\left[a_2(t) + a_2^\dagger(t)\right] + \sqrt{\Gamma}b_{\mathrm{in}}(t). \tag{25b}$$

The above formulation closely resembles Eqs. (17a) and (17b). In an optomechanical cavity coherently driven at one frequency, the up- and down-converted photons (described by δa in two-mode systems (Eq. 22a), and a_2 in three-mode systems (Eq. 25a)) and phonons (δb in Eq. (22b), b in Eq. (25b)) are generated through spontaneous processes, assisted by the vacuum and thermal fluctuations.

3.2 Translationally invariant waveguides

The derivation of the Hamiltonian formulation of opto-acoustic coupling in waveguides is given in detail in Section 3.2 of chapter "Theoretical formalisms for stimulated Brillouin scattering" by Wolff et al. Here, we only briefly revisit this formulation, as we use a slightly different normalization scheme—one that should allow a direct comparison with the cavity optomechanical formulation in Section 3.1.

We begin, as in the cavity systems, with decomposing the electric and acoustic displacement fields $\boldsymbol{E}$ and $\boldsymbol{U}$ into eigenmodes of the optical and acoustic wave equations, respectively, in the form of wave-like solutions

$$\boldsymbol{E}(\boldsymbol{r},t) = \sum_n \int_{-\infty}^{\infty} \frac{\mathrm{d}k}{\sqrt{2\pi}} \tilde{a}_{nk}(t)\exp\left(\mathrm{i}kz - \mathrm{i}\omega_{nk}t\right)\boldsymbol{e}^{(nk)}(x,y) + \mathrm{h.c.}, \tag{26a}$$

$$\boldsymbol{U}(\boldsymbol{r},t) = \sum_m \int_{-\infty}^{\infty} \frac{\mathrm{d}q}{\sqrt{2\pi}} \tilde{b}_{mq}(t)\exp\left(\mathrm{i}qz - \mathrm{i}\Omega_{mq}t\right)\boldsymbol{u}^{(mq)}(x,y) + \mathrm{h.c.} \tag{26b}$$

The two fields can be equivalently described in the momentum space through the optical envelopes $a_n(z, t)$:

$$a_n(z,t) = \int_{-\infty}^{\infty} \frac{\mathrm{d}k}{\sqrt{2\pi}} \exp\left[\mathrm{i}(k-k_n)z - \mathrm{i}\omega_n t\right]\tilde{a}_{nk}(t), \tag{27}$$

and analogously defined acoustic envelopes $b_m(z, t)$. The fields are then quantized, promoting envelopes $\tilde{a}_{nk}$, $\widetilde{b}_{mq}$, $a_n(z)$, and $b_m(z)$ to the photon and photon creation and annihilation operators, defined either in real or momentum space. In the former representation, photon and photon $c = a, b$ operators fulfill

$$[c_m(z,t), c_n^\dagger(z', t)] = \delta_{mn}\delta(z - z'). \tag{28}$$

The operator $\tilde{a}_{nk}^\dagger\tilde{a}_{nk}$ and its transforms $a_n^\dagger(z)a_n(z)$ take on a convenient interpretation of the *photon number density operators* in momentum and real spaces, respectively. This normalization scheme is thus often referred to as *number-density normalization.*

The choice of the expansion of electric and acoustic fields given in Eq. (26) dictates the normalization conditions for the eigenmodes[f]:

$$\int \mathrm{d}\boldsymbol{r}_\perp \varepsilon_0 \varepsilon(\boldsymbol{r}_\perp)|\boldsymbol{e}^{(nk)}(\boldsymbol{r}_\perp)|^2 = \frac{\hbar\omega_{nk}}{2}, \tag{30a}$$

$$\Omega_{mq}^2 \int \mathrm{d}\boldsymbol{r}_\perp \rho(\boldsymbol{r}_\perp)|\boldsymbol{u}^{(mq)}(\boldsymbol{r}_\perp)|^2 = \frac{\hbar\Omega_{mq}}{2}. \tag{30b}$$

We can then express the Hamiltonian of the optical and acoustic modes as

$$\begin{aligned}\mathcal{H}_0^{(\mathrm{OM,wg})} &= \hbar\sum_n \int_{-\infty}^{\infty} \mathrm{d}k\, \omega_{nk}\tilde{a}_{nk}^\dagger\tilde{a}_{nk} + \hbar\sum_m\int_{-\infty}^{\infty}\mathrm{d}q\Omega_{mq}\widetilde{b}_{mq}^\dagger\widetilde{b}_{mq}\\ &\approx \int_{-\infty}^{\infty}\mathrm{d}z\,\hbar\left[\sum_n a_n^\dagger(z)\hat{\omega}_n a_n(z) + \sum_m b_m^\dagger(z)\hat{\Omega}_m b_m(z)\right],\end{aligned} \tag{31a}$$

---

[f] We note that the choice of this convention for the decomposition of the $\boldsymbol{E}$ and $\boldsymbol{U}$ fields is dictated by the correspondence with the formulation widely accepted in the cavity optomechanics community. A similar formulation, historically established in the waveguide optics community, found, for example, in Sipe and Steel (2016), reads:

$$\boldsymbol{E}(\boldsymbol{r},t) = \sum_n \int_{-\infty}^{\infty}\frac{\mathrm{d}k}{\sqrt{2\pi}}\sqrt{\frac{\hbar\omega_{nk}}{2}}\tilde{a}_{nk}(t)\exp(\mathrm{i}kz - \mathrm{i}\omega_{nk}t)\boldsymbol{e}^{(nk)}(x,y) + \mathrm{h.c.}, \tag{29a}$$

$$\boldsymbol{U}(\boldsymbol{r},t) = \sum_m \int_{-\infty}^{\infty}\frac{\mathrm{d}q}{\sqrt{2\pi}}\sqrt{\frac{\hbar\Omega_{mq}}{2}}\widetilde{b}mq(t)\exp(\mathrm{i}qz - \mathrm{i}\Omega_{mq}t)\boldsymbol{u}^{(mq)}(x,y) + \mathrm{h.c.}, \tag{29b}$$

and the right-hand side of the normalization conditions in Eqs. (30) is now equal to 1. Alternatively, one could propose formulations built around redefining the commutators of spatial or momentum operators—see the discussion in chapter "Theoretical formalisms for stimulated Brillouin scattering" by Wolff et al., or Van Laer et al. (2016).

where $\hat{\omega}_n = \omega_{nk_n} - i\partial_z v_g^{(n)}(k_n)$ and $\hat{\Omega}_m = \Omega_{mq_m} - i\partial_z v_g^{(m)}(q_m)$ represent the mode dispersion around the centers of the bands k_n and q_m, using group velocities $v_g^{(n/m)}$.

The opto-acoustic interaction follows then the description given in Section 3.2 of chapter "Theoretical formalisms for stimulated Brillouin scattering" by Wolff et al., which lists both its Hamiltonian formulation, and the Heisenberg equations for the dynamics of $a_n(z, t)$ and $b_m(z, t)$ operators. Here, we repeat these equations in the simplified form, mirroring the dynamical equations of three-mode interaction explored given in Eqs. (24):

$$\left(\frac{\partial}{\partial t} + v_p \frac{\partial}{\partial z}\right) a_p(z,t) = -i\omega_p a_p(z,t) + i\widetilde{g}_0\, a_s(z,t) b(z,t), \tag{32a}$$

$$\left(\frac{\partial}{\partial t} + v_s \frac{\partial}{\partial z}\right) a_s(z,t) = -i\omega_s a_s(z,t) + i\widetilde{g}_0\, a_p(z,t) [b(z,t)]^\dagger, \tag{32b}$$

$$\left(\frac{\partial}{\partial t} + v_b \frac{\partial}{\partial z}\right) b(z,t) = -i\Omega b + i\widetilde{g}_0\, a_p(z,t) [a_s(z,t)]^\dagger, \tag{32c}$$

with $v_{p/s} \equiv v_g^{(p/s)}(k_{p/s})$ and $v_b \equiv v_g^{(b)}(q_b)$. The *single-photon Brillouin coupling coefficient* $\widetilde{g}_0$ (with units $[\widetilde{g}_0] = \mathrm{s}^{-1} \cdot \mathrm{m}^{1/2}$)[g] is related to the Brillouin gain G_B (with $[G_B] = (\mathrm{W \cdot m})^{-1}$) through (Wiederhecker et al., 2019)

$$G_B = 4 \frac{|\widetilde{g}_0|^2}{v_p v_s \Gamma \hbar \omega_p} \frac{(\Gamma/2)^2}{(\Gamma/2)^2 + (\omega_p - \omega_s - \Omega)^2}, \tag{33}$$

with $\omega = \omega_p \approx \omega_s$. Brillouin gain describes the effective coupling between the pump and Stokes fields in a steady state, and the limit of large acoustic damping, as discussed in detail in Section 4.1 in chapter "Theoretical formalisms for stimulated Brillouin scattering" by Wolff et al.

For completeness, as in the case of cavity optomechanics (see Section 3.1.3), these equations should be completed by adding terms describing their interaction with the environment, including loss ($- a_{p/s}\kappa_{p/s}/2$ and $- b\Gamma/2$), as well as the noise and the control fields ($a_{p/s,\mathrm{in}}\sqrt{\kappa_{p/s,\mathrm{ex}}}$ and $b_{\mathrm{in}}\sqrt{\Gamma}$) (Drummond and Corney, 2001).

In typical applications of *stimulated* Brillouin scattering (Brillouin amplification, acoustic memory), optical and acoustic fields can be approximated by their coherent amplitudes ($\alpha_{p/s} = \langle a_{p/s} \rangle$, $\beta = \langle b \rangle$). Since the noise terms describing vacuum and thermal fluctuations are entirely incoherent

[g] The units of $\widetilde{g}_0$ are $\mathrm{m}^{-1} \cdot \mathrm{s}^{1/2}$ when the operators $a_{p/s}(z, t)$ and $b(z, t)$ are flux- (Van Laer et al., 2016) instead of number-density normalized as here.

(see Eqs. 19 and 20), they do not directly contribute to the dynamics of the coherent fields, and the corresponding terms are dropped from the classical Heisenberg-Langevin equations. On the other hand, the noise terms are critical for the accurate description of the Schawlow-Townes linewidth of the Brillouin lasers (Section 4.3), Brillouin cooling (Section 4.1), or the role of thermal fluctuations on the onset of Brillouin scattering (Boyd et al., 1990).

3.2.1 Linearizing the interaction

In a typical three-mode Brillouin setup, one of the optical fields (dubbed the *pump*) is driven strongly, while the other either emerges spontaneously, or is seeded by a much weaker input into the *probe* mode. In either case, the change to the amplitude of the pump mode is mostly determined by the optical losses, with little contribution from the nonlinear interaction. This allows us to linearize the opto-acoustic waveguide Hamiltonian and the corresponding Heisenberg–Langevin equations (see Eqs. 32) by replacing the operators describing the pump mode with the coherent amplitude $\alpha_p(z,t) = \overline{\alpha}_p \exp\left[-i\omega_p t - \kappa_p z/(2v_p)\right]$. In this picture, $\overline{\alpha}_p$ is determined solely by the amplitude of the driving field.

The situation is different for a two-mode intramodal forward configuration, in which we cannot simply replace the only optical envelope with a coherent amplitude—instead, we need to artificially separate the components of optical mode a_p oscillating at the pump frequency, and the sidebands frequency-shifted from it by Ω_{0q_0}. This is a scenario similar to that found in two-mode optomechanics, and is typically resolved by separating a_p into a complex function $\alpha_p(z, t)$, and fluctuation operator $\delta a_p(z, t)$ carrying all the sideband contributions. In this picture, the Brillouin interaction, linearized around $\alpha_p(z, t)$, governs the interaction between $\delta a_p(z, t)$ and $b(z, t)$.

3.3 Linking waveguide and cavity dynamics

3.3.1 Mean-field transition

The dynamics of both waveguides and cavities results from a fundamental three-wave-mixing interaction. Therefore, we can connect their key parameters and physics rigorously. To do so, we perform a so-called *mean-field transition* which averages out the spatial dimension and transforms the space- and time-dependent equations describing a waveguide into the time-dependent equations describing a high-finesse cavity. The transition averages out the spatial variation occurring over one cavity roundtrip (Fig. 3).

In Van Laer et al. (2016), the authors define unitless mean-field envelope operators,[h] which effectively average the fields along the roundtrip length L of a cavity

$$\overline{d}(t) = \frac{1}{\sqrt{L}} \int_0^L \mathrm{d}z \; d(z,t), \tag{34}$$

for the three modes $d = a_p$ (pump), a_s (Stokes), and b (acoustics) involved in the Brillouin interaction. This approach holds rigorously when the spatially averaged fields change slowly over L. Its validity is thus subject to two effects: variation of the local nonlinear interaction strength, and the decay of the fields away from their respective inputs. Both conditions are discussed at length in Van Laer et al. (2016). Here, we only note a counterintuitive observation which extends the validity of this framework: the case of a very short *spatial* acoustic decay length $L \ll 1/\alpha_b = v_b/\Gamma$ does not break this approximation. Indeed, due to the small acoustic velocity v_b, the spatial derivative in Eq. (32c) can be dropped, meaning that the acoustic field does not propagate in space, and is only driven by a local force $\propto a_s^\dagger a_p \propto \overline{a}_s^\dagger \overline{a}_p$.

The resulting coupled ODEs governing the mean-field dynamics are

$$\frac{\mathrm{d}}{\mathrm{d}t}\overline{a}_p = ig_0 \overline{a}_s \overline{b} - \left(\frac{\kappa_p}{2} - i\Delta_p\right)\overline{a}_p, \tag{35a}$$

$$\frac{\mathrm{d}}{\mathrm{d}t}\overline{a}_s = ig_0 \overline{a}_p \overline{b}^\dagger - \left(\frac{\kappa_s}{2} - i\Delta_s\right)\overline{a}_s, \tag{35b}$$

$$\frac{\mathrm{d}}{\mathrm{d}t}\overline{b} = ig_0 \overline{a}_s^\dagger \overline{a}_p - \left(\frac{\Gamma}{2} - i\Delta_b\right)\overline{b}, \tag{35c}$$

with the cavity single-photon coupling parameter g_0 explicitly related to the Brillouin gain G_B (Eq. 33) and the single-photon Brillouin coupling $\widetilde{g}_0$ used in Eqs. (32) via

$$g_0^2 = \frac{\widetilde{g}_0^2}{L} = \frac{v_p v_s \Gamma \hbar \omega_p}{4L} G_B, \tag{36}$$

consistently with Eq. (33).

This mean-field transition was initially made directly on the dynamical equations (Van Laer et al., 2016). An alternative approach where the transition is made on the Hamiltonian is sketched in Safavi-Naeini et al. (2019).

[h] The waveguide operators are flux-normalized with units $s^{-1/2}$ in Van Laer et al. (2016), while they are number-density normalized with units $m^{-1/2}$ here. This difference is accommodated by using the prefactor $L^{-1/2}$ in Eq. (34). The number-density normalization used here also leads to a factor $v_{g,p} v_{g,s} v_{g,b}$ being dropped from Eq. (36) with respect to the corresponding equation derived in Van Laer et al. (2016).

3.3.2 Spatial evolution of the system in steady state

In another limit, we can remove the *temporal* dependence from the Heisenberg equations (32) describing a waveguide, assuming that all the modes of the system have reached their steady states. This approach similarly allows us to simplify the behavior down to a set of three equations reminiscent of the three-mode OM, with a spatial coordinate now replacing the temporal one:

$$\begin{aligned} \frac{\mathrm{d}}{\mathrm{d}z} a_p &= i\widetilde{g}_0 a_s b - \left(\frac{\alpha_p}{2} - i\widetilde{\Delta}_p\right) a_p, \\ \frac{\mathrm{d}}{\mathrm{d}z} a_s &= i\widetilde{g}_0 a_p b^\dagger - \left(\frac{\alpha_s}{2} - i\widetilde{\Delta}_s\right) a_s, \\ \frac{\mathrm{d}}{\mathrm{d}z} b &= i\widetilde{g}_0 a_s^\dagger a_p - \left(\frac{\alpha_b}{2} - i\widetilde{\Delta}_b\right) b, \end{aligned} \tag{37}$$

Wavenumber detunings are represented by $\widetilde{\Delta}_{p/s/b}$.

3.3.3 Waveguides as the limit of optomechanical arrays

Another interesting link between the theoretical formalisms of waveguides and cavities treats a waveguide as the continuum limit of a discrete array of optomechanical cavities (Rakich and Marquardt, 2018). In this model, the interaction can be both local:

$$a_j^\dagger a_j b_j + \text{h.c.} \longrightarrow [a(z)]^\dagger a(z) b(z) + \text{h.c.}, \tag{38}$$

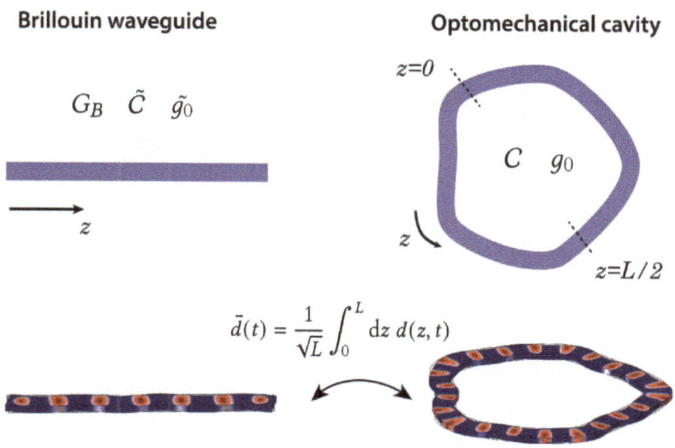

Fig. 3 The physical effects and figures of merit of optomechanical waveguides can be linked rigorously to those of high-finesse cavities (Van Laer et al., 2016). This link is made through a mean-field transition which averages out the spatial variation of the fields over a cavity roundtrip.

and nonlocal, supporting a collection of terms proportional to the products of spatial derivatives of $a(z)$ and $b(z)$, for example,

$$[a(z)]^{\dagger}a(z)\partial_z b(z) + \text{h.c., or } \partial_z[a(z)]^{\dagger}a(z)b(z) + \text{h.c.,} \tag{39}$$

contributing to the spatial evolution of the acoustic and optical excitations.

This formulation allows one to draw direct connections to the effects explored in discrete systems, including the formation of solitons, topological response, or long-range phonon dynamics.

3.4 Characteristic properties of optomechanical waveguides and cavities

3.4.1 Hierarchy of dissipations

The coupled dynamical equations we discuss above appear difficult to solve analytically, whether expressed through quantum-mechanical operators, or classical, complex envelopes, in either the full or the linearized picture. However, both the communities investigating cavity optomechanics and Brillouin scattering have converged on a similar approach for solving them, building on the separation of characteristic timescales for the optical and acoustic fields. Specifically, one of these fields can be considered *fast*, reacting nearly instantaneously to its driving force. It is then formally solved, approximated, and plugged back into the dynamical equations governing the other, *slow* field. The choice of the *fast* and *slow* fields is governed by the immediacy of the response of fields to the forces, or the optical (κ) and acoustic (Γ) decay rates, with the *fast* field being the one with the largest temporal loss. Interestingly, these fundamental parameters turn out to be mostly inverted between the optomechanical cavities and Brillouin systems.

In cavity optomechanics, we nearly always find that the hierarchy of temporal decay rates follows $\kappa > \Gamma$. Thus, from the above prescription, the *fast* optical fields are solved for formally, approximated, and plugged into the equations of motion of the acoustic field. This feedback of the optical cavity onto the mechanical mode is not instantaneous, with the retardation determined by the lifetime of photons in the cavity κ^{-1}. If the delay in feedback is comparable to the mechanical oscillation period, the out-of-phase feedback begins to effectively damp, or amplify the mechanical motion, modifying its acoustic dissipation rate $\Gamma \rightarrow \Gamma + \Gamma_{\text{opt}}$. Through this effect—a result of the *dynamical aspect of the optomechanical backaction*—the acoustic mode can be cooled ($\Gamma_{\text{opt}} > 0$; see Section 4.1 for examples and detailed discussion), amplified ($\Gamma_{\text{opt}} < 0$), or even exhibit dynamical instability and acoustic or phonon lasing ($\Gamma + \Gamma_{\text{opt}} \rightarrow 0$; Section 4.3.1). Such a system can also support optical phenomena enabled by the long coherence of the acoustic mode, such as EIT (Weis et al., 2010) (Section 4.2).

The same hierarchy of temporal dissipation rates was achieved in several specific systems originating from the Brillouin community in which either one of the decay rates was engineered to reach $\kappa > \Gamma$. For example, in ultra-high-Q microresonators, acoustic losses were reduced by harnessing the lower-frequency and lower-loss acoustic modes mediating the forward intermodal Brillouin scattering (Bahl et al., 2012b). On the other hand, the majority of systems studied in the Brillouin community follow the inverse temporal hierarchy: $\kappa < \Gamma$. In this regime, the local or mean acoustic field is adiabatically eliminated, leading to phonon-induced modification of the dissipation and frequency of the Stokes field, to amplification, and eventually lasing of the Stokes field (see Section 4.3).

Additionally, since in Brillouin scattering the fields evolve in space, the optical and acoustic *spatial* decay rates $\alpha_{p/s} = \kappa_{p/s}/v_{p/s}$ and $\alpha_b = \Gamma/v_b$ should play a role in assigning the *fast* and *slow* characteristics in the steady-state regime (Van Laer et al., 2016). Since all of the reported waveguides follow $\alpha_{p/s} \ll \alpha_b$, in the steady-state formulation, the acoustic field is always fast and local. Thus, the Brillouin backaction induces amplification or loss of the optical field along the waveguide. This is the spatial analog of feedback-induced damping or heating. This homogeneity in the hierarchy is mostly due to the many orders of magnitude contrast of the optical and acoustic group velocities $v_{p/s} \gg v_b$. Thus, the hierarchy could potentially be inverted in systems exhibiting low-frequency flexural acoustic modes with $\Omega \propto q^2$, mediating backward Brillouin scattering, and with the losses reduced by low temperatures.

3.4.2 Cooperativities

Linearized optomechanical cavities are naturally characterized by an *effective cooperativity* (Aspelmeyer et al., 2014a)

$$C = \frac{4g^2}{\kappa\Gamma}, \tag{40}$$

which quantifies the efficiency of the optomechanical driving and readout. For example, in a generic optomechanical cavity with $\Omega > \kappa$ (the so-called *resolved sideband* regime), backaction-induced phonon amplification and cooling rates Γ_{opt} are proportional to C, and the phonon lasing threshold is reached at $C = 1$. Furthermore, in optomechanical cavities exhibiting induced transparency, both the depth and the width of transmission window depend on the cooperativity as $(1 + C^{-1})^{-1}$ and $1 + C$, respectively (Weis et al., 2010), see Section 4.2.

As shown in the mean-field treatment connecting waveguides and cavities (Van Laer et al., 2016), we can use the same definition of the cooperativity and, owing to the symmetry of the linearized Brillouin equations, expand it to accommodate different dissipation hierarchies. Specifically, for $\Gamma < \kappa$ and $\Gamma > \kappa$, $C = 1$ would define the thresholds for the lasing in the acoustic and Stokes modes, respectively.

We can also map this definition to a spatial evolution picture, introducing a spatial cooperativity $\widetilde{C} = 4\widetilde{g}^2/(\alpha_s\alpha_b)$ which again carries the same physical interpretation. For Brillouin waveguides, where typically $\alpha_b \gg \alpha_s$, this $\widetilde{C}$ defines the strength of the phonon-assisted attenuation or gain of the Stokes field.

4. Optomechanical phenomena in waveguides and cavities

Building on the converging characteristics of optomechanical and Brillouin systems (Section 2), and the theoretical language used to describe them (Section 3), we now discuss several fundamental phenomena which illustrate the convergence of cavity optomechanics and Brillouin scattering.

4.1 Spontaneous effects

First, we discuss several *spontaneous* effects, occurring when the Brillouin or optomechanical systems are optically driven at only one frequency, and the up- or downshifting of light is induced by the vacuum or thermal fluctuations. This discussion draws from the formulation of the linearized equations of motion for the optomechanical cavities (Section 3.1) and waveguides (Section 3.2), in which we assume that one of the optical modes establishes a coherent state with amplitude α, coupling the remaining modes or fluctuations with an effective coupling $g = g_0|\alpha|$.

4.1.1 Cooling and amplification in cavities

The linearized optomechanical Hamiltonian given in Eq. (23) dictates how the coherent light in the optically driven mode is downconverted to Stokes $(\omega_L - \Omega)$ and upconverted to anti-Stokes $(\omega_L + \Omega)$ frequencies, creating and annihilating phonons. The Stokes and anti-Stokes photons can then either

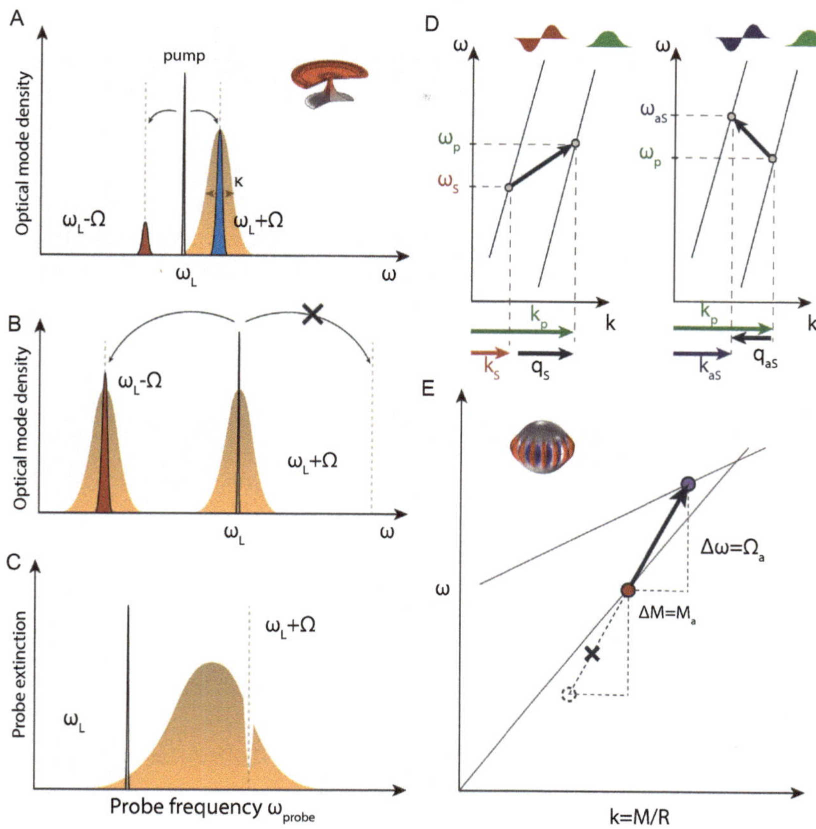

Fig. 4 Optomechanical phenomena identified in waveguides and cavities. (A,B) Cooling and amplification can be implemented in (A) two-mode or (B) three-mode optomechanical cavities by harnessing the contrast in the density of states for the Stokes and anti-Stokes photons engineered by the response of the cavity. (C) In Optomechanically Induced Transparency (see Section 4.2), the extinction of a weak probe beam exhibits a narrow transparency windows centered at $\omega_{\text{probe}} = \omega_L + \Omega$. (D,E) These systems can support similar asymmetry between the density of states for the Stokes and anti-Stokes emission. (D) In waveguides, forward intermodal Brillouin scattering can be used to simultaneously control two different acoustic modes: spontaneous Stokes scattering amplifies the population of Stokes phonons, while the anti-Stokes scattering cools down the counterpropagating, anti-Stokes phonons. (E) In microresonators using optical and acoustic WGMs, the phase-matching condition is replaced by the conservation of azimuthal number, enabling selective realization of the Stokes or anti-Stokes scattering, and nonreciprocal response (see Section 4.2).

populate the same optical cavity mode as the pump (see a schematic in Fig. 4A), or couple to another optical mode a_2 (see Fig. 4B). Engineering these rates (traditionally denoted as Γ_- and Γ_+ for the Stokes and anti-Stokes processes, respectively) enables a number of spontaneous effects, including *cooling*, or *amplification* of the mechanical degree of freedom. It also defines a convenient formulation of the threshold for dynamical instabilities of the system, manifesting as acoustic lasing.

The expressions for the rates $\Gamma_\pm$ can be derived using classical or quantum-mechanical formalisms. In the former, we consider the expectation values of the fluctuation operators $\langle\delta a\rangle$ and $\langle\delta b\rangle$, to see how the optomechanical backaction will change the complex susceptibility of the mechanical mode, modifying both its resonant frequency and dissipation rate (Weis et al., 2010). Starting from Eq. (22a), and dropping the noise terms for simplicity, we can formally solve

$$\begin{aligned}\delta a(t) &= ig\int_0^t \mathrm{d}t' \exp\left[-\left(\frac{\kappa}{2}+i\Delta'\right)(t-t')\right]\left[\delta b(t')+\delta b^\dagger(t')\right] \\ &\approx ig\left[\delta b(t)\chi(\Delta'+\Omega)+\delta b^\dagger(t)\chi(\Delta'-\Omega)\right],\end{aligned} \tag{41}$$

where we performed the Born-Markov approximation, and introduced a spectrum of the optical cavity $\chi(\omega) = (-i\omega - \kappa)^{-1}$ decoupled from mechanical vibrations, in a frame rotating with ω_L. Plugging this solution into Eq. (22b), we find that the optomechanical feedback introduces a correction to the acoustic decay

$$\Gamma_{\text{opt}} = g^2\mathrm{Re}[\chi(\Delta'+\Omega)-\chi(\Delta'-\Omega)] \tag{42a}$$

$$= \underbrace{\frac{g^2\kappa}{(\Delta'+\Omega)^2+(\kappa/2)^2}}_{\Gamma_-} - \underbrace{\frac{g^2\kappa}{(\Delta'-\Omega)^2+(\kappa/2)^2}}_{\Gamma_+}, \tag{42b}$$

and frequency

$$\delta\Omega = g^2\mathrm{Im}[\chi(\Delta'+\Omega)-\chi(\Delta'-\Omega)] \tag{43a}$$

$$= g^2\left[\frac{-\Delta'-\Omega}{(\Delta'+\Omega)^2+(\kappa/2)^2} - \frac{-\Delta'+\Omega}{(\Delta'-\Omega)^2+(\kappa/2)^2}\right]. \tag{43b}$$

Notably, the explicit dependence of Γ_{opt} on κ directly manifests our earlier observation that the modification of the acoustic decay rate Γ_{opt} is due to the retarded optical feedback, and is a *dynamical* optomechanical feedback effect. Conversely, the *optical spring effect*—a frequency shift $\delta\Omega$—would be nonzero for $\kappa = 0$, and is thus dubbed as a *static feedback effect*.

By expanding the above derivation to account for the noise terms, we can similarly derive the effective dynamics of the incoherent phonon population $n_b = \langle\delta b^\dagger\delta b\rangle$, and identify the laser-like rate equation

$$\frac{\mathrm{d}}{\mathrm{d}t}n_b = -(\Gamma + \Gamma_-)n_b + \Gamma_+(n_b + 1) + \Gamma n_b^{\mathrm{th}}. \tag{44}$$

Its steady-state solution

$$n_{b,ss} = \frac{\Gamma_+ + \Gamma n_b^{\mathrm{th}}}{\Gamma + \Gamma_{\mathrm{opt}}}, \tag{45}$$

readily demonstrates that positive Γ_{opt} can lead to cooling of the mechanical mode, negative Γ_{opt} induces amplification, and we define the *threshold for dynamical instabilities* or *phonon lasing* as $\Gamma + \Gamma_{\mathrm{opt}} = 0$. In a two-mode optomechanical system operating in the resolved sideband limit $\Omega > \kappa$, and for optimally detuned pump, these rates can be expressed through the cooperativity C (see Eq. 40). Specifically, for the blue-detuned coherent driving $\Gamma_{\mathrm{opt}} \approx \Gamma_+ = C\,\Gamma$, leading to the amplification of the mechanical mode with a rate $(C - 1)\Gamma$, and phonon lasing at $C = 1$. Conversely, an optimally red-detuned laser can induce cooling at a rate $\Gamma_{\mathrm{opt}} \approx \Gamma_- = C\,\Gamma$.

Equation (42) introduces the correction to the phonon decay rate Γ_{opt} in terms of the spectra of the optical system. This concept can be directly extended to multimodal systems, with more complex χ. For example, in three-mode optomechanics (see Fig. 4B), the contrast between Γ_+ and Γ_- can be even larger, if only one of the sidebands is tuned to a narrow optical mode. However, an experimental realization of such a system requires careful matching of the intermode spacing to the mechanical frequency.

Optomechanical cooling has been experimentally realized in multiple systems, including optomechanical crystals, and high-frequency toroidal microresonators. Historically, the first two demonstrations of the near-ground state sideband laser cooling in these two-mode optomechanical systems were reported by Chan et al. (2011) and Verhagen et al. (2012), respectively.[i] The former considered a radiation pressure optomechanical coupling between the 3.7 GHz mechanical mode and a near-IR optical cavity mode with $\kappa/2\pi \sim 500$ MHz (sideband resolution $\Omega/\kappa \sim 7.5$); the cryostat precooled system (at 20 K, or $n_b^{\mathrm{th}} \approx 100$) was optically cooled until the incoherent phonon population reached $n_b \sim 0.85$. The latter realized coupling between an optical WGM and an RBM ($\Omega/2\pi = 78$ MHz and $\Omega/\kappa = 11$). The toroidal resonator was precooled by a cryostat to $n_b^{\mathrm{th}} \approx 200$, and optically cooled to $n_b \approx 1.7$.

Cooling has also been observed in resonators in which the interacting photons exchange momentum by coupling to an acoustic WGM, with

[i] As well as in the microwave regime by Teufel et al. (2011).

the modes following the $\Gamma \ll \kappa$ dissipation hierarchy. This scenario, identified historically as a case of *Brillouin cooling*, was explored by Bahl et al. (2012a), who coupled an acoustic WGM (at frequency $\Omega/2\pi \sim 95$ MHz, and azimuthal number M) and two optical WGMs ($\omega_{1/2}$, $m_{1/2}$) in an ultra-high-optical-Q microcavity ($Q > 10^8$). The WGMs were selected to fulfill the frequency ($\omega_1 + \Omega = \omega_2$) and phase-matching (expressed by the relationship between the azimuthal numbers of the WGMs $m_1 + M = m_2$) conditions, and the interaction was due to the photoelastic/electrostrictive effect. For cooling, the pump laser was set to resonantly drive the lower-frequency mode $\omega_L = \omega_1$, matching the spontaneously generated anti-Stokes light to the upper-frequency mode ω_2 (Fig. 4E). A complementary Stokes process was suppressed by ensuring that there were no optical modes frequency- ($\omega_L - \Omega$) and phase-matched ($m_1 - M$) to the Stokes transition. The anti-Stokes transitions depleted the thermal population of the acoustic WGM from room temperature to about 19 K, simultaneously broadening the intrinsic phonon linewidth from 7.7 kHz to about 118 kHz (consistently with $C \approx 15$). The same asymmetry can be also used to implement phonon amplification, by setting the pump to the higher-frequency mode ω_2, to promote the Stokes scattering.

Conversely, the demonstration of strong optomechanical amplification requires much smaller cooperativities, limited by the onset of lasing at $C = 1$, and is of less interest for a majority of studies. However, one recently proposed exception is its application to *molecular optomechanics*, which described the coupling between optical cavities and Raman transitions in molecules which they host. In those systems, optomechanical backaction enables dramatic enhancement of the intrinsically weak sensitivity of the Raman spectroscopy (Lombardi et al., 2018).

4.1.2 Cooling and amplification in waveguides

Cooling and amplification of the mechanical vibrations can be also realized in waveguides with a cavity optomechanics-like hierarchy of dissipations $\Gamma \ll \kappa$, as discussed in Section 3.4.1. An example of such a setup was reported in Kittlaus et al. (2017), where the authors realized inter-modal Brillouin scattering between two *different* forward-propagating optical modes, mediated by a few-GHz, low-quality ($Q_b \sim 500$) acoustic mode. In Otterstrom et al. (2018a), the authors then used this platform to demonstrate optomechanical cooling in a waveguide (Fig. 4D). Here, however, the clear parallel to a two-mode cavity optomechanics breaks down on multiple fronts. Most importantly, acoustic modes of a waveguide form a continuum, and the cooling efficiency becomes a spectrum with width explicitly dependent on the waveguide length L. To derive this result, Otterstrom et al. (2018a) embraced a particular Hamiltonian formulation of Brillouin dynamics, in which the optical modes are represented in spatial coordinates, while the acoustic excitation is represented in momentum space. Second, two *different* phonon modes mediate the Stokes and anti-Stokes scattering. In particular, as depicted schematically in Fig. 4D, the forward-propagating *Stokes* phonons mediate the Stokes scattering, and experience amplification, while the backward-propagating *anti-Stokes phonons* experience cooling.

4.2 Coherently driven effects below the threshold

Instead of relying on the *spontaneous* up- and downconversion, one can interrogate opto-acoustic systems by measuring how strong the coherent driving modifies the transmission or reflection of weak *probe* signals. In fact, this is the typical application scenario for Brillouin amplifiers in which a weak probe beam becomes amplified as it propagates along the waveguide, drawing energy from the strong coherent field established by the pump (see Eq. 33 and discussion in Section 4.3.3 of chapter "Theoretical formalisms for stimulated Brillouin scattering" by Wolff et al.)

Another effect which explicitly harnesses the coherences developing in driven opto-acoustic systems is an analog of the electromagnetically induced transparency (EIT). This effect has been observed in optomechanical cavities (Dong et al., 2015; Kim et al., 2015; Safavi-Naeini et al., 2011; Weis et al., 2010), and been referred to as optomechanically induced transparency (OMIT) or Brillouin scattering induced transparency (BSIT), depending on the details of the realization.

Before we discuss these experimental reports, let us briefly review the physics of the induced transparency. A canonical EIT considers the absorption of a weak coherent probe beam by a Λ system spanned by three states: $\{|1\rangle, |2\rangle, |e\rangle\}$, with a pump creating a strong coherence between states $|2\rangle$ and $|e\rangle$. As the probe becomes resonant with transition $|1\rangle \leftrightarrow |e\rangle$, a destructive interference between the pathways $|1\rangle \rightarrow |e\rangle$ and $|1\rangle \rightarrow |e\rangle \rightarrow |2\rangle \rightarrow |e\rangle$ forbids the excitation of $|e\rangle$, rendering the system transparent to the probe. The same effect can be induced in a two-mode opto-acoustic system by engineering the states of the Λ system as $\{|1\rangle = |n, m\rangle, |2\rangle = |n, m+1\rangle, |e\rangle = |n+1, m\rangle\}$ with n photons and m phonons. Under red-detuned pumping ($\omega_L = \omega_1 - \Omega$), the transmission of a weak probe around $\omega_L + \Omega \approx \omega_1$ has a Lorentzian profile with width $\Gamma_{\text{OMIT}} = \Gamma(1 + C)$, and the maximum normalized transmission proportional to $(1 + C^{-1})^{-1}$.

This scenario has been realized in Weis et al. (2010), using an optical WGM mode of a toroidal resonators, coupled to a low-frequency ($\Omega/2\pi \sim 50$ MHz) RBM. This system demonstrated normalized transmission of about 81%, consistent with $C \approx 9$. In Safavi-Naeini et al. (2011), opto-acoustic EIT was observed at room temperature, in an optomechanical crystal with a few-GHz acoustic mode with $Q_b \approx 2 \times 10^3$, with cooperativity up to $C \approx 10$.

Opto-acoustic EIT has been also realized in Brillouin-like setups, where the coupling between two optical modes was mediated by the low-loss *traveling* acoustic waves, supporting the $\kappa \gg \Gamma$ dissipation hierarchy. This was first demonstrated in silica microspheres, by frequency- and phase-matching two optical WGMs and an acoustic WGM (Dong et al., 2015; Kim et al., 2015) in a forward scattering configuration. The opto-acoustic coupling imposed the additional condition of phase-matching (for WGMs: $M + \mathrm{m}_1 = \mathrm{m}_2$), fulfilled only for the copropagating optical modes in a forward scattering configuration. Authors in Dong et al. (2015) and Kim et al. (2015) used this additional requirement to induce a nonreciprocal response of the system: a probe copropagating with the strong pump beam was transmitted through the cavity, while for the counterpropagating optical beams, the interaction was strongly phase-mismatched, and the probe was absorbed.

4.3 Above the threshold

In Sections 3.4.1 and 4.1, we discussed how the hierarchy of dissipation rates is a key characteristic of both waveguides and cavities. Specifically, we focused on the scenario where the Stokes or anti-Stokes photons and phonons are created at rates far lower than κ and Γ. Here, we discuss a complementary scheme, where the rate of inelastic scattering processes approaches either one of these rates, leading to a build-up of excitations in the system, and eventually a laser-like threshold behavior.

Optomechanical systems can support both optical and acoustic lasing, depending on the hierarchy of dissipation rates. For a detailed outlook into the fundamental theory of Brillouin lasers, and their physical realization, we direct the reader to chapter "Integrated Brillouin lasers and their applications" by Blumenthal et al. Here, we briefly mention some of the modern implementations of such devices.

As a point of clarification, we note that the term *Brillouin lasers* is traditionally associated with systems in which the lasing threshold is defined by the rate of generation of Stokes photons matching the Stokes decay rate, meaning that it can only occur in systems where $\kappa < \Gamma$. Beyond this optical Brillouin lasing threshold, the Stokes field settles into a coherent state with large amplitude, and exhibits linewidth narrowing. Conversely, if the system exhibits cavity-optomechanics-like hierarchy of dissipation $\kappa > \Gamma$, the lasing threshold occurs when the *phonon* generation rate reaches Γ, and the acoustic field exhibits line narrowing.

4.3.1 Acoustic lasing

In optomechanical cavities, we usually find that the acoustic dissipation rate Γ is well below the optical rate κ, suggesting that as the net phonon generation rate reaches Γ, the system will exhibit *phonon lasing* (also referred to as *sasing* or *mechanical self-oscillation*).

The first experimental observations of this effect were reported in a system driven by photothermal forces comprising a microlever (Metzger and Karrai, 2004), oscillating at approximately 7 kHz ($\Gamma/2\pi \sim 3$ Hz), which served as an end mirror of a Fabry–Pérot cavity, similar to the idealized scenario shown in Fig. 1A.

Soon after, phonon lasing in high-mechanical frequency systems was realized by harnessing the moving boundary coupling in optomechanical microresonators. In Carmon et al. (2005) and Kippenberg et al. (2005) the authors observed and analyzed coupling between modes of an ultra-high-Q resonator with ~10 μm radii, coupled to a number of RBMs with frequencies spanning 5–100 MHz. The study demonstrated that the phonon lasing threshold is inversely proportional to the acoustic mode quality factor Q_b over the entire considered range of frequencies, with the lowest threshold powers around 20 μW. The characteristics of the mechanical oscillations above the threshold were then analyzed in detail in Rokhsari et al. (2006), revealing thermal broadening of their linewidth.

Since these fundamental demonstrations, phonon lasing has been reported in a plethora of other cavity optomechanical systems, including membrane-in-the-middle setups (Chen et al., 2015; Sheng et al., 2020), levitated nanoparticles (Pettit et al., 2019), and arrays of synchronized optomechanical resonators (Gil-Santos et al., 2017).

Acoustic lasing was also realized in more exotic systems: for example, in He et al. (2020), a silica microdisk resonator of 40 μm radius was coated with a few-nanometer-thick superfluid helium film. In this system, the Brillouin interaction between the counterpropagating optical WGMs ($\kappa/2\pi = 284$ MHz) and low-frequency ($\Omega/2\pi \sim 7$ MHz) acoustic waves propagating in superfluid helium ($\Gamma/2\pi \sim 85$ kHz) demonstrated phonon lasing threshold as low as 1.8 μW. Unlike the microresonators in Bahl et al. (2011) or waveguides in Otterstrom et al. (2018b), this setup harnessed the *backward* Brillouin interaction, enabled by the greatly reduced acoustic velocities in the superfluid (few meters per second).

4.3.2 Optical lasing

Optical lasers based on Brillouin scattering have routinely been realized in fiber loops since the 1970s (Hill et al., 1976), by reversing the traditional dissipation hierarchy of cavity optomechanics via the ultra-low loss of an optical fiber. Realizing an *optical laser in microresonators* was harder since it required identifying systems where the spectral spacing of the optical modes would match the acoustic frequency. This challenge was first successfully addressed in 2009 in Grudinin et al. (2009), by coupling optical ultra-high-Q ($\sim 10^9$, $\kappa/2\pi \sim 0.3$ MHz) optical WGMs of a much larger, *millimeter*-sized calcium fluorite resonator to a high-frequency 17.7 GHz acoustic mode with considerable losses ($\Gamma/2\pi \sim 10$ MHz). Soon after, Brillouin lasing in an integrated, silicon wedge microresonator was reported in Lee et al. (2012), paving the way toward realizing extremely narrow, sub-hertz laser linewidths.

Optical Brillouin lasers have been also implemented in several other semi- or fully integrated systems. In Kabakova et al. (2013), authors reported on lasing in a system comprising a few centimeter long, straight *chalcogenide* (As_2S_3) waveguide, with the optical cavity loop closed by coupling the Stokes field to a low-loss fiber. Over the next several years, these systems will be fully integrated into on-chip architectures, for example, as bus-ring *silicon nitride* (Si_3N_4) waveguide resonators (Gundavarapu et al., 2019). The refined design and improved fabrication techniques led to the further, dramatic reduction of the Brillouin laser linewidths toward a sub-hertz regime (Gundavarapu et al., 2019).

5. Summary

The communities of cavity optomechanics and Brillouin scattering are converging on many fronts, sailing past their historical differences. They explore similar, often outright identical, devices and rely on the same type of three-wave-mixing interactions between optical and acoustic modes. The two fields also use increasingly homogeneous theoretical, numerical, and experimental tools. They explore phenomena which can often be directly and quantitatively mapped onto each other. This convergence has already and should continue to enable successful cross-pollination of ideas between the communities. We expect this convergence to have an impact on both fundamental and practical research topics, potentially venturing into emerging quantum technologies and large-scale integrated photonics.

Acknowledgments

M.K.S. acknowledges funding from the Macquarie University Research Fellowship Scheme (MQRF0001036) and the Australian Research Council Discovery Early Career Researcher Award DE220101272. C.G.B. acknowledges support from Australian Research Council Fellowship DE190100318 and UQ ECR Grant 2058737. R.V.L. acknowledges support from the Wallenberg Center for Quantum Technology and from the European Research Council with a Starting Grant.

References

Arcizet, O., Cohadon, P.F., Briant, T., Pinard, M., Heidmann, A., 2006. Radiation-pressure cooling and optomechanical instability of a micromirror. Nature 444 (7115), 71–74. https://doi.org/10.1038/nature05244. https://www.nature.com/articles/nature05244.

Armani, D.K., Kippenberg, T.J., Spillane, S.M., Vahala, K.J., 2003. Ultra-high-Q toroid microcavity on a chip. Nature 421 (6926), 925–928. https://doi.org/10.1038/nature01371. https://www.nature.com/articles/nature01371.

Aspelmeyer, M., Kippenberg, T.J., Marquardt, F., 2014a. Cavity optomechanics. Rev. Mod. Phys. 86 (4), 1391–1452. https://doi.org/10.1103/RevModPhys.86.1391.

Aspelmeyer, M., Kippenberg, T.J., Marquardt, F., 2014b. Cavity Optomechanics: Nano-and Micromechanical Resonators Interacting With Light. Springer, Berlin, Heidelberg, https://doi.org/10.1007/978-3-642-55312-7.

Bahl, G., Zehnpfennig, J., Tomes, M., Carmon, T., 2011. Stimulated optomechanical excitation of surface acoustic waves in a microdevice. Nat. Commun. 2 (1), 1–6. https://doi.org/10.1038/ncomms1412. https://www.nature.com/articles/ncomms1412.

Bahl, G., Tomes, M., Marquardt, F., Carmon, T., 2012a. Observation of spontaneous Brillouin cooling. Nat. Phys. 8 (3), 203–207. https://doi.org/10.1038/nphys2206. https://www.nature.com/articles/nphys2206.

Bahl, G., Tomes, M., Marquardt, F., Carmon, T., 2012b. Observation of spontaneous Brillouin cooling. Nat. Phys. 8 (3), 203–207.

Baker, C., Hease, W., Nguyen, D.-T., Andronico, A., Ducci, S., Leo, G., Favero, I., 2014. Photoelastic coupling in gallium arsenide optomechanical disk resonators. Opt. Express 22 (12), 14072–14086. https://doi.org/10.1364/OE.22.014072. http://www.opticsexpress.org/abstract.cfm?URI=oe-22-12-14072.

Balram, K.C., Davanço, M., Lim, J.Y., Song, J.D., Srinivasan, K., 2014. Moving boundary and photoelastic coupling in GaAs optomechanical resonators. Optica 1 (6), 414–420. https://doi.org/10.1364/OPTICA.1.000414. https://www.osapublishing.org/optica/abstract.cfm?uri=optica-1-6-414.

Boyd, R.W., Rzaˌewski, K., Narum, P., 1990. Noise initiation of stimulated Brillouin scattering. Phys. Rev. A 42 (9), 5514–5521. https://doi.org/10.1103/PhysRevA.42.5514.

Braginsky, V.B., Manukin, A.B., 1967. Ponderomotive effects of electromagnetic radiation. Sov. Phys. JETP 25 (4), 653–655.

Braginsky, V.B., Strigin, S.E., Vyatchanin, S.P., 2001. Parametric oscillatory instability in Fabry-Perot interferometer. Phys. Lett. A 287 (5), 331–338. https://doi.org/10.1016/S0375-9601(01)00510-2.

Brillouin, L., 1922. Diffusion de la lumière et des rayons X par un corps transparent homogène. Ann. Phys. 9 (17), 88–122. https://doi.org/10.1051/anphys/192209170088. https://www.annphys.org/articles/anphys/abs/1922/17/anphys19220917p88/anphys19220917p88.html.

Carmon, T., Rokhsari, H., Yang, L., Kippenberg, T.J., Vahala, K.J., 2005. Temporal behavior of radiation-pressure-induced vibrations of an optical microcavity phonon mode. Phys. Rev. Lett. 94 (22), 223902. https://doi.org/10.1103/PhysRevLett.94.223902.

Chan, J., Alegre, T.P.M., Safavi-Naeini, A.H., Hill, J.T., Krause, A., Gröblacher, S., Aspelmeyer, M., Painter, O., 2011. Laser cooling of a nanomechanical oscillator into its quantum ground state. Nature 478 (7367), 89–92. https://doi.org/10.1038/nature10461. https://www.nature.com/articles/nature10461.

Chan, J., Safavi-Naeini, A.H., Hill, J.T., Meenehan, S., Painter, O., 2012. Optimized optomechanical crystal cavity with acoustic radiation shield. Appl. Phys. Lett. 101 (8), 081115. https://doi.org/10.1063/1.4747726. https://aip.scitation.org/doi/full/10.1063/1.4747726.

Chen, X., Zhao, C., Danilishin, S., Ju, L., Blair, D., Wang, H., Vyatchanin, S.P., Molinelli, C., Kuhn, A., Gras, S., et al., 2015. Observation of three-mode parametric instability. Phys. Rev. A 91 (3), 033832. https://doi.org/10.1103/PhysRevA.91.033832.

Ding, L., Baker, C., Senellart, P., Lemaitre, A., Ducci, S., Leo, G., Favero, I., 2010. High frequency GaAs nano-optomechanical disk resonator. Phys. Rev. Lett. 105 (26), 263903. https://doi.org/10.1103/PhysRevLett.105.263903.

Ding, L., Baker, C., Senellart, P., Lemaitre, A., Ducci, S., Leo, G., Favero, I., 2011. Wavelength-sized GaAs optomechanical resonators with gigahertz frequency. Appl. Phys. Lett. 98 (11), 113108. https://doi.org/10.1063/1.3563711.

Dong, C.-H., Shen, Z., Zou, C.-L., Zhang, Y.-L., Fu, W., Guo, G.-C., 2015. Brillouin-scattering-induced transparency and non-reciprocal light storage. Nat. Commun. 6 (1), 1–6. https://doi.org/10.1038/ncomms7193. https://www.nature.com/articles/ncomms7193.

Dorsel, A., McCullen, J.D., Meystre, P., Vignes, E., Walther, H., 1983. Optical bistability and mirror confinement induced by radiation pressure. Phys. Rev. Lett. 51 (17), 1550. https://doi.org/10.1103/PhysRevLett.51.1550. https://journals.aps.org/prl/abstract/10.1103/PhysRevLett.51.1550.

Drummond, P.D., Corney, J.F., 2001. Quantum noise in optical fibers I Stochastic equations. J. Opt. Soc. Am. B 18 (2), 139. https://doi.org/10.1364/josab.18.000139.

Eggleton, B.J., Poulton, C.G., Pant, R., 2013. Inducing and harnessing stimulated Brillouin scattering in photonic integrated circuits. Adv. Opt. Photonics 5 (4), 536. https://doi.org/10.1364/AOP.5.000536. https://www.osapublishing.org/aop/abstract.cfm?uri=aop-5-4-536.

Eggleton, B.J., Poulton, C.G., Rakich, P.T., Steel, M.J., Bahl, G., 2019. Brillouin integrated photonics. Nat. Photonics 13 (10), 664–677. https://doi.org/10.1038/s41566-019-0498-z. http://www.nature.com/articles/s41566-019-0498-z.

Gil-Santos, E., Labousse, M., Baker, C., Goetschy, A., Hease, W., Gomez, C., Lemaître, A., Leo, G., Ciuti, C., Favero, I., 2017. Light-mediated cascaded locking of multiple nano-optomechanical oscillators. Phys. Rev. Lett. 118 (6), 063605. https://doi.org/10.1103/PhysRevLett.118.063605.

Grudinin, I.S., Matsko, A.B., Maleki, L., 2009. Brillouin lasing with a CaF_2 whispering gallery mode resonator. Phys. Rev. Lett. 102 (4), 043902. https://doi.org/10.1103/PhysRevLett.102.043902.

Gundavarapu, S., Brodnik, G.M., Puckett, M., Huffman, T., Bose, D., Behunin, R., Wu, J., Qiu, T., Pinho, C., Chauhan, N., et al., 2019. Sub-hertz fundamental linewidth photonic integrated Brillouin laser. Nat. Photonics 13 (1), 60–67. https://doi.org/10.1038/s41566-018-0313-2. https://www.nature.com/articles/s41566-018-0313-2.

Han, K., Kim, J., Bahl, G., 2013. Radiation-driven optomechanical pressure sensor. In: Frontiers in Optics 2013, Optical Society of America, p. FTu4C.3. http://www.osapublishing.org/abstract.cfm?URI=FiO-2013-FTu4C.3.

He, X., Harris, G.I., Baker, C.G., Sawadsky, A., Sfendla, Y.L., Sachkou, Y.P., Forstner, S., Bowen, W.P., 2020. Strong optical coupling through superfluid Brillouin lasing. Nat. Phys. 16 (4), 417–421. https://doi.org/10.1038/s41567-020-0785-0. https://www.nature.com/articles/s41567-020-0785-0.

Hill, K.O., Kawasaki, B.S., Johnson, D.C., 1976. cw Brillouin laser. Appl. Phys. Lett. 28 (10), 608–609. https://doi.org/10.1063/1.88583.

Hill, J.T., Safavi-Naeini, A.H., Chan, J., Painter, O., 2012. Coherent optical wavelength conversion via cavity optomechanics. Nat. Commun. 3 (1), 1196. https://doi.org/10.1038/ncomms2201.

Kabakova, I.V., Pant, R., Choi, D.-Y., Debbarma, S., Luther-Davies, B., Madden, S.J., Eggleton, B.J., 2013. Narrow linewidth Brillouin laser based on chalcogenide photonic chip. Opt. Lett. 38 (17), 3208–3211. https://doi.org/10.1364/OL.38.003208. https://opg.optica.org/ol/fulltext.cfm?uri=ol-38-17-3208.

Kim, J., Kuzyk, M.C., Han, K., Wang, H., Bahl, G., 2015. Non-reciprocal Brillouin scattering induced transparency. Nat. Phys. 11 (3), 275–280. https://doi.org/10.1038/nphys3236. https://www.nature.com/articles/nphys3236.

Kipfstuhl, L., Guldner, F., Riedrich-Möller, J., Becher, C., 2014. Modeling of optomechanical coupling in a phoxonic crystal cavity in diamond. Opt. Express 22 (10), 12410–12423. https://doi.org/10.1364/OE.22.012410. https://www.osapublishing.org/oe/abstract.cfm?uri=oe-22-10-12410.

Kippenberg, T.J., Vahala, K.J., 2007. Cavity opto-mechanics. Opt. Express 15 (25), 17172–17205. https://doi.org/10.1364/OE.15.017172. http://www.osapublishing.org/oe/abstract.cfm?URI=oe-15-25-17172.

Kippenberg, T.J., Rokhsari, H., Carmon, T., Scherer, A., Vahala, K.J., 2005. Analysis of radiation-pressure induced mechanical oscillation of an optical microcavity. Phys. Rev. Lett. 95 (3), 033901. https://doi.org/10.1103/PhysRevLett.95.033901.

Kittlaus, E.A., Otterstrom, N.T., Rakich, P.T., 2017. On-chip inter-modal Brillouin scattering. Nat. Commun. 8 (1), 1–9. https://doi.org/10.1038/ncomms15819. https://www.nature.com/articles/ncomms15819.

Kobyakov, A., Sauer, M., Chowdhury, D., 2010. Stimulated Brillouin scattering in optical fibers. Adv. Opt. Photonics 2 (1), 1–59. https://doi.org/10.1364/AOP.2.000001. https://opg.optica.org/aop/abstract.cfm?uri=aop-2-1-1.

Lebedew, P., 1901. Untersuchungen über die druckkräfte des lichtes. Ann. Phys. 311 (11), 433–458. https://doi.org/10.1002/andp.19013111102.

Lee, H., Chen, T., Li, J., Yang, K.Y., Jeon, S., Painter, O., Vahala, K.J., 2012. Chemically etched ultrahigh-Q wedge-resonator on a silicon chip. Nat. Photonics 6 (6), 369–373. https://doi.org/10.1038/nphoton.2012.109. https://www.nature.com/articles/nphoton.2012.109.

Lombardi, A., Schmidt, M.K., Weller, L., Deacon, W.M., Benz, F., de Nijs, B., Aizpurua, J., Baumberg, J.J., 2018. Pulsed molecular optomechanics in plasmonic nanocavities: from nonlinear vibrational instabilities to bond-breaking. Phys. Rev. X 8 (1), 011016. https://doi.org/10.1103/PhysRevX.8.011016.

Mandelstam, L.I., 1926. Light scattering by inhomogeneous media. Zh. Russ. Fiz-Khim. 58, 381–386.

Metzger, C.H., Karrai, K., 2004. Cavity cooling of a microlever. Nature 432 (7020), 1002–1005. https://doi.org/10.1038/nature03118. https://www.nature.com/articles/nature03118.

Nelson, D.F., 1991. Momentum, pseudomomentum, and wave momentum: Toward resolving the Minkowski-Abraham controversy. Phys. Rev. A 44 (6), 3985–3996. https://doi.org/10.1103/PhysRevA.44.3985.

Nichols, E.F., Hull, G.F., 1901. A preliminary communication on the pressure of heat and light radiation. Phys. Rev. (Series I) 13 (5), 307. https://doi.org/10.1103/PhysRevSeriesI.13.307.

Nunnenkamp, A., Sudhir, V., Feofanov, A.K., Roulet, A., Kippenberg, T.J., 2014. Quantum-limited amplification and parametric instability in the reversed dissipation regime of cavity optomechanics. Phys. Rev. Lett. 113 (2), 023604. https://doi.org/10.1103/PhysRevLett.113.023604.

Otterstrom, N.T., Behunin, R.O., Kittlaus, E.A., Rakich, P.T., 2018a. Optomechanical cooling in a continuous system. Phys. Rev. X 8 (4), 041034. https://doi.org/10.1103/PhysRevX.8.041034.

Otterstrom, N.T., Behunin, R.O., Kittlaus, E.A., Wang, Z., Rakich, P.T., 2018b. A silicon Brillouin laser. Science 360 (6393), 1113–1116. https://doi.org/10.1126/science.aar6113.

Pettit, R.M., Ge, W., Kumar, P., Luntz-Martin, D.R., Schultz, J.T., Neukirch, L.P., Bhattacharya, M., Vamivakas, A.N., 2019. An optical tweezer phonon laser. Nat. Photonics 13 (6), 402–405. https://doi.org/10.1038/s41566-019-0395-5. https://www.nature.com/articles/s41566-019-0395-5.

Rakich, P., Marquardt, F., 2018. Quantum theory of continuum optomechanics. New J. Phys. 20 (4), 045005. https://doi.org/10.1088/1367-2630/aaac4f.

Rakich, P.T., Reinke, C., Camacho, R., Davids, P., Wang, Z., 2012. Giant enhancement of stimulated Brillouin scattering in the subwavelength limit. Phys. Rev. X 2 (1), 011008. https://doi.org/10.1103/PhysRevX.2.011008.

Rokhsari, H., Hossein-Zadeh, M., Hajimiri, A., Vahala, K., 2006. Brownian noise in radiation-pressure-driven micromechanical oscillators. Appl. Phys. Lett. 89 (26), 261109. https://doi.org/10.1063/1.2424276.

Safavi-Naeini, A.H., Painter, O., 2010. Design of optomechanical cavities and waveguides on a simultaneous bandgap phononic-photonic crystal slab. Opt. Express 18 (14), 14926–14943. https://doi.org/10.1364/OE.18.014926. https://opg.optica.org/oe/fulltext.cfm?uri=oe-18-14-14926.

Safavi-Naeini, A.H., Alegre, T.P.M., Chan, J., Eichenfield, M., Winger, M., Lin, Q., Hill, J.T., Chang, D.E., Painter, O., 2011. Electromagnetically induced transparency and slow light with optomechanics. Nature 472 (7341), 69–73.

Safavi-Naeini, A.H., Van Thourhout, D., Baets, R., Van Laer, R., 2019. Controlling phonons and photons at the wavelength scale: integrated photonics meets integrated phononics. Optica 6 (2), 213–232. https://doi.org/10.1364/OPTICA.6.000213. https://opg.optica.org/optica/fulltext.cfm?uri=optica-6-2-213.

Sheard, B.S., Gray, M.B., Mow-Lowry, C.M., McClelland, D.E., Whitcomb, S.E., 2004. Observation and characterization of an optical spring. Phys. Rev. A 69 (5), 051801. https://doi.org/10.1103/PhysRevA.69.051801.

Sheng, J., Wei, X., Yang, C., Wu, H., 2020. Self-organized synchronization of phonon lasers. Phys. Rev. Lett. 124 (5), 053604. https://doi.org/10.1103/PhysRevLett.124.053604.

Shin, H., Qiu, W., Jarecki, R., Cox, J.A., Olsson, R.H., Starbuck, A., Wang, Z., Rakich, P.T., 2013. Tailorable stimulated Brillouin scattering in nanoscale silicon waveguides. Nat. Commun. 4, 1944. https://doi.org/10.1038/ncomms2943. https://www.nature.com/articles/ncomms2943.

Sipe, J.E., Steel, M.J., 2016. A Hamiltonian treatment of stimulated Brillouin scattering in nanoscale integrated waveguides. New J. Phys. 18 (4), 045004. https://doi.org/10.1088/1367-2630/18/4/045004.

Teufel, J.D., Donner, T., Li, D., Harlow, J.W., Allman, M.S., Cicak, K., Sirois, A.J., Whittaker, J.D., Lehnert, K.W., Simmonds, R.W., 2011. Sideband cooling of micromechanical motion to the quantum ground state. Nature 475 (7356), 359–363. https://doi.org/10.1038/nature10261. http://www.nature.com/nature/journal/v475/n7356/abs/nature10261.html.

Thompson, J.D., Zwickl, B.M., Jayich, A.M., Marquardt, F., Girvin, S.M., Harris, J.G.E., 2008. Strong dispersive coupling of a high-finesse cavity to a micromechanical membrane. Nature 452 (7183), 72–75. https://doi.org/10.1038/nature06715. http://www.nature.com/nature/journal/v452/n7183/abs/nature06715.html.

Van Laer, R., 2016. Light-sound interaction in nanoscale silicon waveguides, Chapter 2. arXiv:1605.09645. PhD thesis. http://arxiv.org/abs/1605.09645.

Van Laer, R., Bazin, A., Kuyken, B., Baets, R., Van Thourhout, D., 2015a. Net on-chip Brillouin gain based on suspended silicon nanowires. New J. Phys. 17 (11), 115005. https://doi.org/10.1088/1367-2630/17/11/115005.

Van Laer, R., Kuyken, B., Van Thourhout, D., Baets, R., 2015b. Interaction between light and highly confined hypersound in a silicon photonic nanowire. Nat. Photonics 9 (3), 199–203. https://doi.org/10.1038/nphoton.2015.11. http://www.nature.com/articles/nphoton.2015.11.

Van Laer, R., Baets, R., Van Thourhout, D., 2016. Unifying Brillouin scattering and cavity optomechanics. Phys. Rev. A 93 (5), 053828. https://doi.org/10.1103/PhysRevA.93.053828.

Verhagen, E., Deléglise, S., Weis, S., Schliesser, A., Kippenberg, T.J., 2012. Quantum-coherent coupling of a mechanical oscillator to an optical cavity mode. Nature 482 (7383), 63–67. https://doi.org/10.1038/nature10787. https://www.nature.com/articles/nature10787.

Weis, S., Rivière, R., Deléglise, S., Gavartin, E., Arcizet, O., Schliesser, A., Kippenberg, T.J., 2010. Optomechanically induced transparency. Science 330 (6010), 1520–1523. https://doi.org/10.1126/science.1195596.

Wiederhecker, G.S., Dainese, P., Mayer Alegre, T.P., 2019. Brillouin optomechanics in nanophotonic structures. APL Photonics 4 (7), 071101. https://doi.org/10.1063/1.5088169.

Wolff, C., Steel, M.J., Eggleton, B.J., Poulton, C.G., 2015. Stimulated Brillouin scattering in integrated photonic waveguides: forces, scattering mechanisms, and coupled-mode analysis. Phys. Rev. A 92 (1), 013836. https://doi.org/10.1103/PhysRevA.92.013836.

Zhu, Z., Gauthier, D.J., Boyd, R.W., 2007. Stored light in an optical fiber via stimulated Brillouin scattering. Science 318 (5857), 1748–1750. https://doi.org/10.1126/science.1149066.

CHAPTER FOUR

Brillouin scattering in photonic crystal fibers

Paulo Dainese[a,*], Jean-Charles Beugnot[b], Fan Yang[c], and Meng Pang[d]
[a]Corning Research and Development Corporation, Corning, NY, United States
[b]FEMTO-ST Institute, UMR 6174 CNRS, Universite Bourgogne Franche Comte, Besançon, France
[c]Ecole Polytechnique Fédérale de Lausanne (EPFL), Lausanne, Switzerland
[d]Shanghai Institute of Optics and Fine Mechanics, Chinese Academy of Sciences, Shanghai, China
*Corresponding author: e-mail address: dainesep@corning.com

Contents

1. Introduction

Photonic crystal fibers (PCFs) consist of a center core region, solid or hollow, surrounded by a periodic microstructured cladding (Russell, 2003). Proper engineering of the cladding enables light guidance through a modified total internal reflection mechanism in solid cores or through photonic bandgap in hollow-core PCFs. These fibers have created new opportunities to study light matter interaction, particularly involving nonlinear optical effects. In solid-core PCFs, optical confinement in a small mode area leads to high optical intensities while low propagation loss leads to long interaction distances, two conditions that strongly enhance nonlinear effects. Furthermore, engineering of the microstructure allows precise control of

Semiconductors and Semimetals, Volume 109
ISSN 0080-8784
https://doi.org/10.1016/bs.semsem.2022.04.006

dispersion and polarization properties. Motivated by these attributes, various phenomena such as soliton propagation (Skryabin et al., 2003), four-wave mixing (Wadsworth et al., 2004), Brillouin scattering (Dainese et al., 2006a), and supercontinuum generation (Dudley et al., 2006) have been extensively investigated in solid-core PCFs. Similarly, hollow-core fibers can also transmit high intensities over long distances and, in particular, can be filled with gases enabling extensive studies of Raman (Benabid et al., 2002a) and Brillouin (Yang et al., 2020) scattering.

Early experiments on Brillouin scattering in PCFs have leveraged these two attributes, high nonlinearity and long interaction length. In 1999, Broderick et al. reported the first experiments on Brillouin scattering in PCFs, and observed stimulated scattering at lower power levels compared to standard fibers due to smaller mode effective area (Fig. 1A). Following this observation, several papers have investigated Brillouin scattering in PCFs, both in backward and forward directions, as well as Brillouin lasing and slow-light applications (De Matos et al., 2004; Furusawa et al., 2006; Harun et al., 2008; Lee et al., 2002; Nasir et al., 2009; Shibata et al., 2006; Yang et al., 2008; Zou et al., 2003).

Beyond small mode area and long interaction length, interest in Brillouin scattering in particular has been further escalated due to another fundamental aspect: while the complex microstructure of the core and cladding impacts strongly the properties of guided *optical* modes, they also have a significant effect on the behavior of the *mechanical* modes. A variety of spatial, polarization and spectral properties of different mechanical modes supported by solid- or hollow-core fibers has led to a series of demonstrations over the last decade or so, some of which are illustrated in Fig. 1. These demonstrations have established PCFs as an effective platform to investigate novel fundamental aspects of the interaction between light and mechanical waves as well as to explore potential applications, in some ways paralleling developments in microcavities and integrated waveguides. In this chapter, the basic properties of Brillouin scattering in PCFs are reviewed along with a discussion of the experimental results they have enabled. In Section 2, we focus on the spatial and polarization properties of mechanical modes in solid-core fibers and how they impact Brillouin scattering, followed by a discussion of time-domain and coherent effects and their applications for mode-locking and storage. In Section 3, we present a discussion on the mechanical modes and associated effects in hollow-core fibers. Finally, we conclude in Section 4.

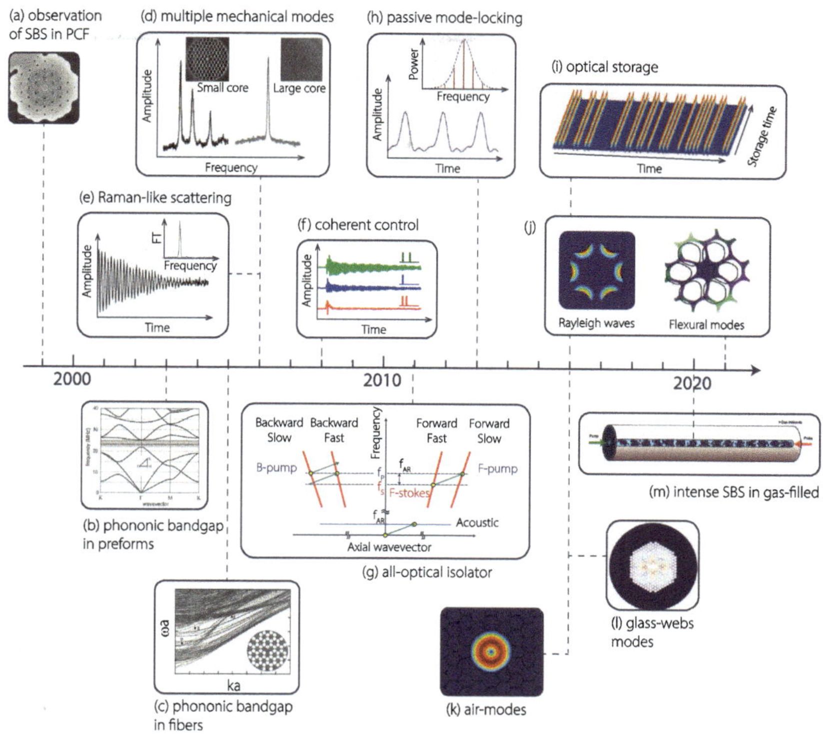

Fig. 1 Timeline highlighting demonstrations of fundamental aspects and new applications of Brillouin scattering in photonic crystal fibers: (A) first observation of Brillouin scattering in PCF (Broderick et al., 1999); confinement of mechanical waves through phononinc bandgap in (B) preforms (Russell et al., 2003) and in (C) fibers (Laude et al., 2005); (D) observation of multiple mechanical modes in backward Brillouin scattering (Dainese et al., 2006a); (E) single-resonance forward scattering in a process that resembles Raman scattering (Dainese et al., 2006b); (F) coherent control of a mechanical mode through a series of timed optical pulses (Wiederhecker et al., 2008); (G) all-optical isolator leveraging polarization dependence of Brillouin scattering involving torsional mechanical modes (Kang et al., 2011); (H) high-frequency passive mode-locking enabled by a 1.8 GHz breathing radial mechanical (Kang et al., 2013); (I) all-optical storage enabled by solitons strongly bound by optomechanical interaction (Pang et al., 2016); (J) observation of scattering arising from surface (Tchahame et al., 2016) and flexural (Jarschel et al., 2021) mechanical modes; Brillouin scattering in hollow-core fibers arising from pressure waves (K) in air (Renninger et al., 2016a) and (L) from mechanical modes of the glass webs surrounding the core (Renninger et al., 2016b); (M) strong stimulated backward Brillouin scattering in gas-filled hollow-core fibers (Yang et al., 2020).

2. Optomechanical interaction in solid-core PCFs

In the early in 2000s, it was recognized that a microstructured cladding would impact the behavior of the mechanical modes supported by PCFs. The idea that a *phononic* bandgap could be created by proper engineering of the cladding structure and used to confine mechanical energy in the core was explored initially in preforms through simulations and experimentally (Russell et al., 2003), and later in fibers through simulations (Laude et al., 2005) (Fig. 1B and C). It was later understood that, to some extent, mechanical modes can be confined in the core region even if a phononic bandgap is not present, resembling modes in a small glass rod suspended in air. This aspect started to manifest itself in a series of experiments on both forward (Beugnot et al., 2007b; Dainese et al., 2006b; Elser et al., 2006; Kang et al., 2009) and backward (Beugnot et al., 2007a; Dainese et al., 2006a; McElhenny et al., 2008; Minardo et al., 2008) Brillouin scattering in PCFs. In Dainese et al. (2006a), a high intensity laser launched into a silica fiber with small core diameter (~1.2 μm diameter) and with high air-fill fraction back-scattered into multiple Brillouin-shifted tones (Fig. 1D). This is in contrast to a single tone observed when the same experiment was performed in a large core (~9.3 μm diameter) and low air-fill fraction. The later resembles typical results obtained in bulk silica glass, with a single back-scattered peak centered at around 11.2 GHz arising due to the interaction of the light beam with a purely longitudinal mechanical wave. At around the same time, *forward* Brillouin scattering also turned out to exhibit intriguing results (Beugnot et al., 2007b; Dainese et al., 2006b; Elser et al., 2006; Kang et al., 2009). While in standard solid fibers, the forward Brillouin spectrum exhibits 10s of peaks with frequencies ranging from 10s to 100s of MHz (Shelby et al., 1985), the result in small core PCFs showed only a few or even a single mechanical resonance at much higher (GHz) frequencies (Fig. 1E).

To understand the physics behind these observations, we review the characteristics of Brillouin scattering in PCFs with different geometries (Figs. 2 and 3). Details of the formalism we followed are given in Wiederhecker et al. (2019). First, the fundamental optical mode is calculated numerically at the input wavelength of 1.55 μm. With its propagation constant at hand (β), we then compute the mechanical modes with propagation constant (q) that satisfy the phase-matching condition, which for back-scattering is $q \approx \beta - (-\beta) = 2\beta$ and for forward it is $q \approx \beta - \beta = 0$

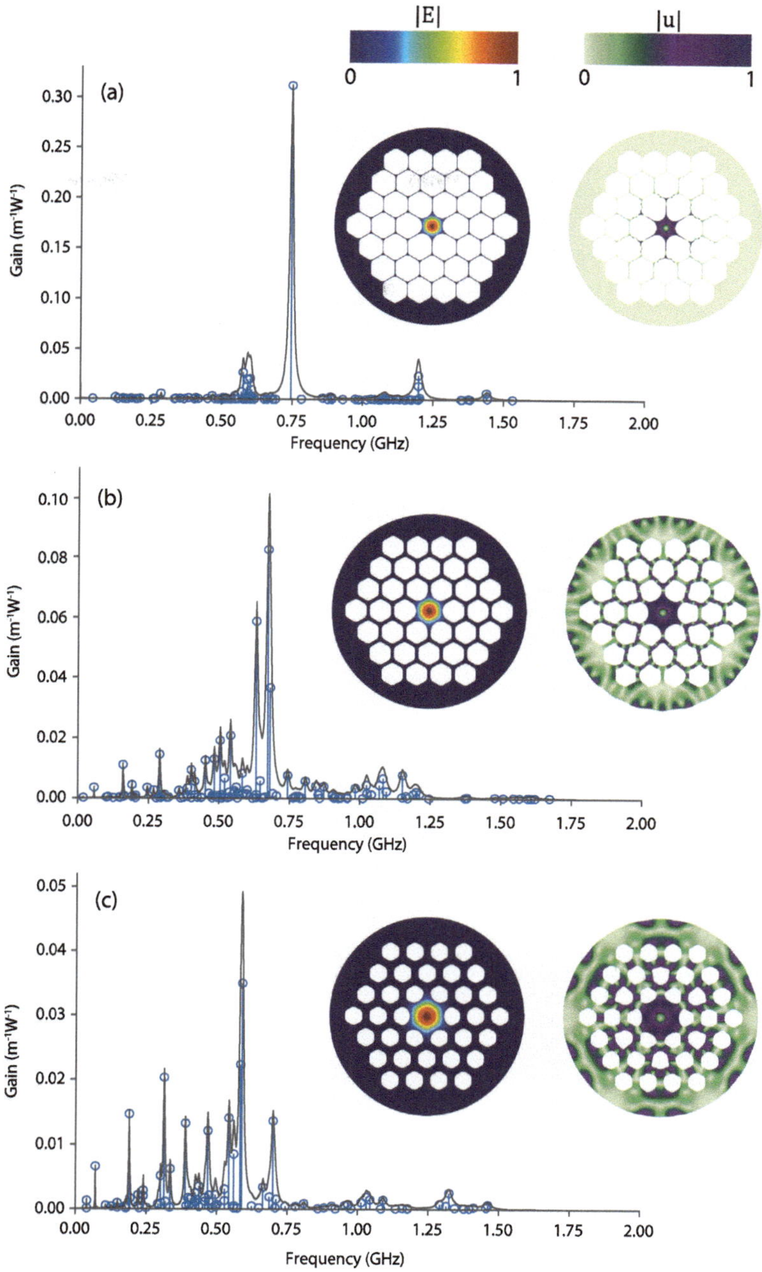

Fig. 2 Forward Brillouin spectrum calculated for three photonic crystal fibers having the same hexagonal cladding structure, pitch (5 μm) and core radius (2 μm) but with decreasing air-fill fraction from top to bottom. Fibers with high air-fill and small core diameter exhibit almost single-peaked spectrum. In each plot, the *inset* shows the optical mode profile (intensity) and the dominant mechanical mode profile (norm of displacement vector). *Color scale* is normalized from 0 to 1. The spectrum is calculated following the formalism in Wiederhecker et al. (2019), using optical and mechanical material properties of fused silica, 1.55 μm wavelength, and assuming a quality factor of 100 for all mechanical modes. Only modes with gain larger than 10^{-4} $m^{-1}W^{-1}$ are included in the plots.

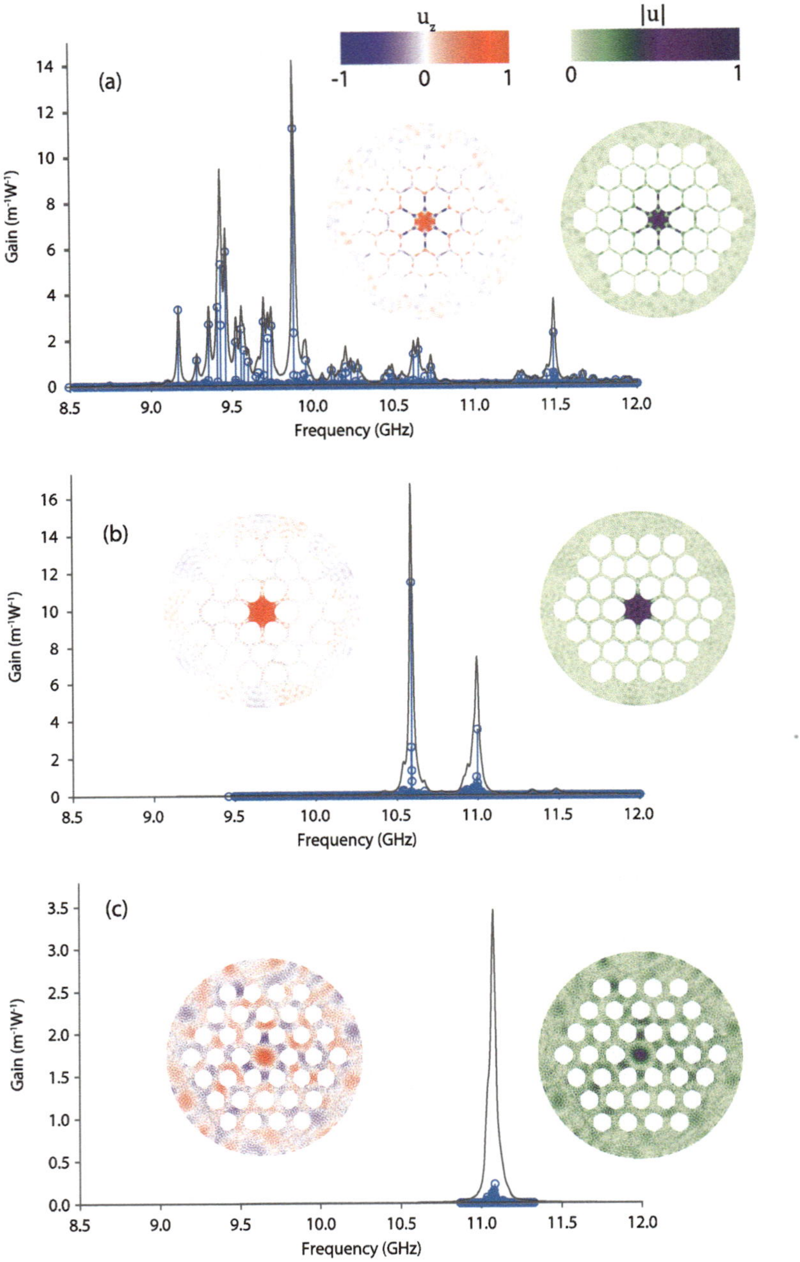

Fig. 3 Backward Brillouin spectrum calculated for three photonic crystal fibers having the same hexagonal cladding structure, pitch (5 μm) and with increasing core diameter (1 μm, 2 μm, 6 μm) and decreasing air-fill fraction from top to bottom. Different than the forward case, fibers with high air-fill and small core diameter now exhibit multipeaked Brillouin spectrum. In each plot, the *insets* show both the longitudinal displacement field u_z (*red-blue color scale*, normalized from -1 to 1) and the norm of the total displacement field $|u|$ (*green-purple color scale*, normalized from 0 to 1). The spectrum is calculated following the formalism in Wiederhecker et al. (2019), using optical and mechanical material properties of fused silica, 1.55 μm wavelength, and assuming a quality factor of 1000 for all mechanical modes. Only modes with gain larger than 10^{-6} $m^{-1}W^{-1}$ are included in the plots.

(this is only an approximation because strictly speaking the two optical waves have slightly different frequencies, shifted from one another by the mechanical frequency, leading to different wavenumbers). We also consider here that both optical waves propagate in the same mode and with the same polarization. The phase-matching condition determines the frequency of each mechanical mode interacting with light. The strength of the interaction is then determined by spatial overlap integrals involving the optical and mechanical mode fields. Specifically, each mechanical mode interacts with the optical mode through the photoelastic and moving-boundary mechanisms (or conversely through electrostriction or radiation pressure forces), both contributing to a total Brillouin gain coefficient for that particular mode. The overall Brillouin gain spectrum is then obtained by appropriately summing up the contributions from all phase-matched mechanical modes.

2.1 Forward Brillouin spectrum

Fig. 2A–C shows the *forward* Brillouin gain spectra for three fibers with the same hexagonal cladding structure, pitch, and core size but with decreasing air-fill fraction from top to bottom. The open circles indicate the set of discrete gain coefficients calculated for the phase-matched mechanical modes, while the continuous curve is the total summed response (even though each mode contributes with a Lorentzian response (Wiederhecker et al., 2019), the sum may yield different combined line shapes). In all cases, both input and Stokes fields propagate in the fundamental mode (and same polarization state).

The optical mode is confined into the core region in all three cases (see insets), although it is noticeable that reducing the air holes slightly increases the core region and consequently the mode area. The main effect, however, is in the structure of the mechanical modes, and as a consequence, in the resulting Brillouin spectrum. One can clearly see that in the highest air-fill structure of Figs. 2A, there is mostly a single mechanical resonance that interacts efficiently with light. The inset shows the norm of the mechanical displacement field ($|u|$) for this dominant mode. This mode couples very weakly with the surrounding microstructure cladding, and resembles the mode of a free-standing rod of glass (Auld and Green, 1974; Thurston, 1978; Waldron, 1969; Wiederhecker et al., 2019). Intuitively, the surface of the core is mostly free, connected to the cladding only through very thin glass webs. Indeed, the mechanical displacement of this dominant mode points primarily in the radial direction (*breathing* mode), similar to the fundamental mode of the axial–radial (AR) family in a glass rod (Wiederhecker et al., 2019). Note that in circular rods, *azimuthally* symmetric modes at cut-off $q = 0$ have only one nonzero component of the displacement field: torsional modes ($u_\phi = 0$), longitudinal modes ($u_z = 0$) or radial modes ($u_r = 0$), the latter being responsible for forward scattering in circular fibers (Auld and Green, 1974; Waldron, 1969).

One might wonder why only the fundamental mode of the AR-like family interacts strongly with light. This is even more intriguing when we consider that in conventional solid fibers, forward Brillouin scattering is also well explained by mechanical modes in a rod and does involve many high-order AR modes (Shelby et al., 1985). The origin of this different behavior is related to the confinement of mechanical modes in the core of the PCFs. Higher order modes exhibit nodes *in the core region* that effectively reduce the overlap with the optical fields. Physically this simply means positive and negative strain regions partially cancel out the net permittivity perturbation (conversely, the mechanical wave generated by the electrostrictive force resembles more closely the fundamental mechanical mode (Beugnot and Laude, 2012)). In conventional fibers, only the optical mode is confined in the core region, while the transverse mechanical modes are spread out in the entire fiber (core and cladding). As a result, many high-order modes do not exhibit any nodes in the core region, and no such cancelation effect occurs (of course they have nodes in the cladding but that does not alter the overlap with the optical field). Significant interaction occurs for many modes up until a high enough order that starts to exhibit nodes in the core.

As the air-fill fraction decreases in Fig. 2B and C, it is clear that the dominant mode couples more strongly with the cladding structure. With the energy distributed throughout a larger area, the strength of the interaction with the optical mode is reduced, and this is reflected in the lower Brillouin gain compared to Fig. 2A. The weaker transverse confinement also lowers the vibration frequencies (in a rod, the frequency is inversely proportional to the radius (Jarschel et al., 2018; Kang et al., 2008; Shelby et al., 1985; Wiederhecker et al., 2019)). It is also noticeable the appearance of several lower frequency peaks. These are predominantly cladding modes that now extend into the core and have nonnegligible interaction with the optical mode. One can imagine that bringing the air-fill to zero simply turns the structure into a solid glass rod, and eventually the spectrum would have resonances that are fully determined by the cladding diameter, as is the case in standard fibers (Shelby et al., 1985). The influence of air-fill fraction on forward Raman-like scattering has been investigated experimentally and numerically in Brenn et al. (2009), while frequency-selective excitation of modes in forward scattering was explored in Stiller et al. (2011).

This single-frequency process observed in small core and high air-fill fraction PCFs resembles the process of Raman scattering (Dainese et al., 2006b; Kang et al., 2009), involving however transverse mechanical phonons at GHz frequencies instead of optical phonons at much higher frequencies (see insets of Fig. 1C). Such reduction in the number of mechanical resonances has been explored as means of reducing noise in quantum platforms (Elser et al., 2006), and has also led to demonstrations of highly coherent excitation of mechanical vibrations in the core with a properly timed sequence of pulses (Fig. 1F). Later, this single-frequency resonance was leveraged in a series of experiments to create mode-locking in fiber lasers at GHz frequencies (Fig. 1H), as well as in experiments where optical solitons were bound by the mechanical potential to create all-optical bit storage (Fig. 1I). We discuss these aspects in Section 2.4. Finally, it has been shown that these high-frequency resonances exhibit strong temperature sensitivity, about 10 times larger than previously reported in conventional single-mode fibers, making this potentially interesting for applications in optical fiber sensors (Carry et al., 2011).

2.2 Backward Brillouin scattering

In Fig. 3, we show the *backward* Brillouin gain spectra for three fibers with the same hexagonal cladding structure and pitch, but with increasing core size and decreasing air-fill fraction from top to bottom. Similarly to the previous *forward* case in Fig. 2, from top to bottom the structure supports more and more mechanical modes with a significant fraction of energy in the core region. Again, this is not only because larger cores support more mechanical modes as one would expect for a large rod suspended in air, but also due to the fact that the cladding structure couples strongly with the core with resulting modes that penetrate into the core region. Surprisingly though, the spectrum evolves in the opposite way as observed in the forward case: here the top spectrum is highly multipeaked, despite supporting fewer modes. In the middle and bottom cases, despite with increasingly more modes in the core region, the spectrum evolves toward a narrower frequency range, eventually becoming single peaked. Note that in both Fig. 3B and C, even though the resulting spectra have well defined peaks, the density of mechanical modes is so high that each scattering peak is formed by coherently adding the contribution of many modes with frequencies falling within the Brillouin linewidth. These modes become increasingly more longitudinal in nature, resembling eventually a continuum of modes in bulk. As their frequencies deviate from the phase matching at 11.2 GHz, rapid transverse variations reduce the overall photoelastic effect, resulting in a single-peaked Brillouin spectrum.

The discussion presented here considers a pure silica glass fiber. Interestingly, when dopants are added to modify the refractive index profile in the core region, multipeaked backward Brillouin spectrum has also been observed (Zou et al., 2003, 2012), even for low air-fill fraction fibers. An index-guidance mechanism has been attributed to explain the existence of these peaks, in the sense that the doping profile creates a center region with lower acoustic velocity compared to the surrounding silica region, similarly to conventional fibers engineered to control Brillouin scattering (Kobyakov et al., 2010). This mechanism however has not been strong enough to confine the transverse mechanical modes involved in forward scattering. Fibers made entirely of other materials, such as chalcogenide, have also been investigated. Multiple Brillouin resonances in backward configuration have been observed with very large Brillouin gain, mainly due to large photoelastic coefficients (Cheng et al., 2012; Cherif et al., 2012; Fatome et al., 2009; Fortier et al., 2008; Shahi et al., 2009; Soltanian et al., 2015).

Another aspect that is different in the backward and forward cases is the nature of the mechanical modes. While in the forward case the mechanical modes are mostly transverse in nature, the *dominant* mode in backward scattering have predominantly axial displacement field u_z. This behavior can be seen in the insets of Fig. 3, where the field profiles of u_z closely resembles the profile of the total displacement field ($|u|$). It is also noticeable the presence of nodes in u_z for the smaller core fiber (Fig. 3A), which results from strong coupling between shear and longitudinal waves at the glass-air boundary as well as strong coupling with the surrounding glass webs. These nodes reduce the overall elasto-optic effect, since physically they decrease the average index perturbation or in other words, these nodes reduce the overlap integrals between optical fields and strain fields (Wiederhecker et al., 2019). The consequence of this is that in fibers with very small core size and high air-fill fraction, the Brillouin gain is actually less than what is predicted by the Smith's formula (Smith, 1972), which considers simply the scaling of the optical intensity optical with core size. This has been observed experimentally in Dainese et al. (2006a).

The properties of the mechanical modes discussed so far have been explored in several applications. The high sensitivity of the mechanical frequencies to the geometrical parameters can be leveraged to enhance or suppress Brillouin scattering. Fibers with dimensions varying along the length leads to a broader Brillouin spectrum and higher stimulated threshold (Poletti et al., 2007; Stiller et al., 2012). Tapered PCFs were fabricated to control the Brillouin line shape and effective Brillouin threshold reduction of 6 dB (Beugnot et al., 2014; Tchahame et al., 2015; Woodward et al., 2014). Large mode area PCFs is another route to reduce Brillouin efficiency (Limpert et al., 2003). Distributed Brillouin scattering measurement on PCFs enables characterization of the structural irregularities during the fabrication process (Beugnot et al., 2007a; Hong et al., 2010; Stiller et al., 2010). Another area of applications is in sensing where it has been demonstrated that different peaks in the backward Brillouin spectrum exhibit different dependencies with temperature and strain (Liang et al., 2013; Zou et al., 2004, 2012). In 2012, Zou et al. (2012) demonstrate a Brillouin spectrum with five different resonances (9.7 GHz, 10.0 GHz, 10.3 GHz, 10.5 GHz, and 10.9 GHz) with temperature and strain coefficient equal to 0.99 MHz/°C and 0.038 MHz/$\mu\epsilon$. Recently, stimulated backward Brillouin scattering has been studied in helical PCFs (Zeng et al., 2021). Interestingly, right-circularly polarized light is back-scattered into left-circularly polarized light as required by angular momentum conservation and, moreover, maintains its polarization state stable along this type of fiber.

2.3 Surface effects

The glass-air core boundary has important implications to Brillouin scattering in PCFs. In all-solid fibers the dominant underlying physical mechanism is the photoelastic effect (pe) in which the dielectric permittivity is disturbed by the elastic strain. In contrast, the contribution from the moving-boundary effect (mb), in which the optical boundary is distorted by the mechanical wave, can be significant given the large index contrast between air and glass. In terms of optical forces, these two effects are directly related to excitation of mechanical waves by electrostriction or by radiation forces (Wiederhecker et al., 2019).

To illustrate this point, we show in Fig. 4A the contribution from each effect in a forward scattering configuration. The plot specifically shows the force densities, $F_{mb} = \int f_{mb} dl$ and $F_{pe} = \int f_{pe} da$, which are, respectively, the line and area overlap integrals representing the mb and pe effects

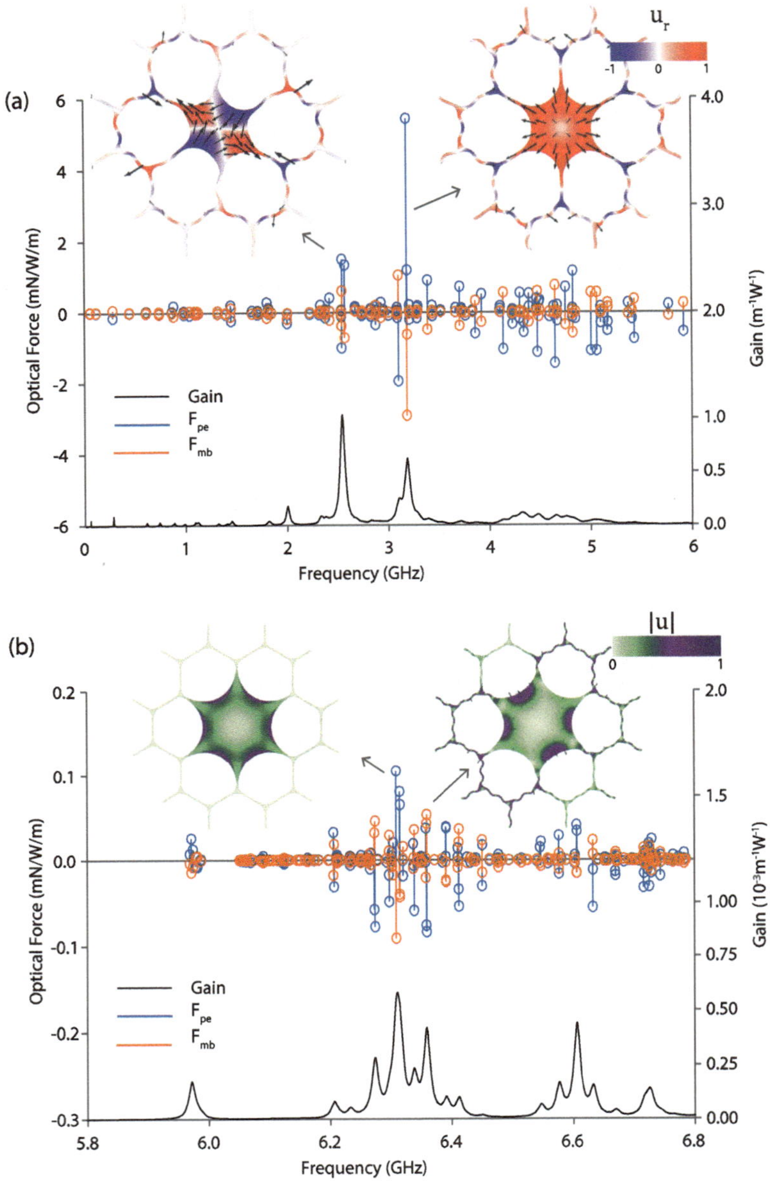

Fig. 4 Surface effects in PCFs: in (A) photoelastic (pe) and moving-boundary (mb) contributions to the forward Brillouin spectrum in a fiber with 2 μm pitch and 1 μm core diameter. The *insets* show the radial displacement component u_r for two dominant modes, one resembling fundamental axial–radial and torsional-radial modes of a rod (Auld and Green, 1974; Thurston, 1978; Waldron, 1969) (*color scale* is normalized from −1 to 1). In (B), backward Brillouin scattering arising from surface waves at around 6 GHz. Again, photoelastic (pe) and moving-boundary (mb) contributions are shown. The *insets* show the norm of total displacement field u. we plot the spectrum is calculated following the formalism in Wiederhecker et al. (2019), using optical and mechanical material properties of fused silica, 1.55 μm wavelength, and assuming a quality factor of 100 for forward and 1000 for backward cases.

(Wiederhecker et al., 2019). The corresponding Brillouin spectrum is plotted in the same figure, on the secondary right axis. This was calculated for a fiber with 1 μm core diameter and 2 μm pitch. One can see that for most modes, both pe and mb effects are comparable in magnitude. Moreover, for some modes these two forces have opposite signs, meaning that the force excitation due to radiation pressure counteracts the electrostrictive forces (Florez et al., 2016). In particular, one can see that the mode with the highest electrostrictive force at around 3.2 GHz also has a strong opposing radiation pressure force. This is a *breathing* mode as can be observed in the figure inset, similar to the dominant peak previously discussed for the larger core fiber (2 μm) in Fig. 2. However, this counter-action of the two forces reduces the total Brillouin gain, and this mode is no longer dominant. Instead, a mode around 2.6 GHz dominates the spectrum. This is a mode resembling torsional-radial modes with azimuthal index 2 in circular rods (Dainese et al., 2006a; Shelby et al., 1985; Thurston, 1978; Waldron, 1969), which for this particular case has both mb and pe forces constructively interfering. Given their spatial profile, these torsional-radial modes also cause polarization coupling. For example, in Kang et al. (2010) orthogonally polarized modes of a birefringent PCF interacts with torsional-radial mechanical modes, leading an effect known as forward stimulated interpolarization scattering. This effect was later explored to demonstrate an all-optical isolator in PCFs (Kang et al., 2011) (Fig. 1G).

A second aspect associated with the free glass-air boundary is the existence of Rayleigh surface waves (Fig. 1J). In silica surfaces, they propagate with a velocity of approximately 3410 m/s (Beugnot et al., 2014; Thurston, 1978), lower than both longitudinal and shear waves. They were predicted to interact with 1550 nm light at around 6–7 GHz (Dainese et al., 2006a) and experimentally observed recently (Tchahame et al., 2016). Fig. 4B shows the backward spectrum in the same fiber as in Fig. 4A, with scattering peaks at around 6 GHz. The inset shows two examples of mechanical modes in this region, in which most energy is concentrated near the air-glass boundary. As a final example of how the spatial properties of the mechanical modes in PCF impact Brillouin scattering, Jarschel et al. (2021) demonstrated optomechanical coupling between the fundamental and the first high-order optical mode (LP11-like) through flexural modes in the PCFs (Fig. 1J). These modes cause the suspended core to bounce back and forth in air, supported only by the glass webs. From an optomechanical point of view, the large moving-boundary effect associated with the flexural mode induces high Brillouin gain. Experimentally, however, low intermodal efficiency was observed due to significant air damping of flexural modes (Jarschel et al., 2021).

2.4 Coherence and time-domain effects

In the previous section, we discussed the spatial characteristics of the mechanical modes and how they impact the resulting Brillouin spectrum in PCFs for both forward and backward configuration. Here, we now focus on coherent effects and time-domain manifestation of the optomechanical interaction involving such modes. We start with the phenomenon of Raman-like scattering, in which a single mechanical resonance predominantly interacts with light in the forward configuration (Dainese et al., 2006b; Kang et al., 2009; Shelby et al., 1985). This effect in PCFs is strongly enhanced compared to conventional solid fibers for two main reasons: first, as already discussed in the introduction, high optical intensity can be achieved in PCFs with very small core. Second, strong confinement of the mechanical resonance in the fiber core leads to strong overall optomechanical coupling. This was also illustrated in Fig. 2A for the *breathing* modes in high air-filling fibers. Overall, the strength of forward Brillouin effects in PCFs can be approximately two orders of magnitude higher than that in the conventional single-mode fiber. Another unique aspect is that the frequency can be very high, on the order of several GHz, and can be controlled by adjusting the core size.

Since the forward Brillouin scattering process can lead to the strong coupling and energy transfer between two optical frequencies propagating in the same direction, cascaded effects can easily take place. Fig. 5A illustrates the dispersion diagram of the cascaded forward Brillouin process. Multiple optical components equally spaced in frequency are coupled together through a *single* optically driven mechanical mode. As a result, the phase-locking between these optical components is automatically established, forming a comb-like structure. In the time domain, these phase-locked comb lines correspond to a train of optical pulses with a repetition rate equal to the mechanical frequency (Fig. 5).

A time-domain picture can be useful to further understand the physics of the optomechanical interaction. A train of short optical pulses with a repetition rate close to the mechanical resonance frequency can coherently drive a transverse mechanical wave in the PCF core and, as a back-action, the mechanical wave shifts the central frequency of the optical pulses to the longer-wavelength side. This red-shift of the optical pulses results from the energy transfer from the optical to the mechanical wave, which can also be understood as the energy flow from the higher-frequency components (pump photons) to the lower frequency ones (Stokes photons). We had previously discussed that the mechanical modes involved in forward Brillouin

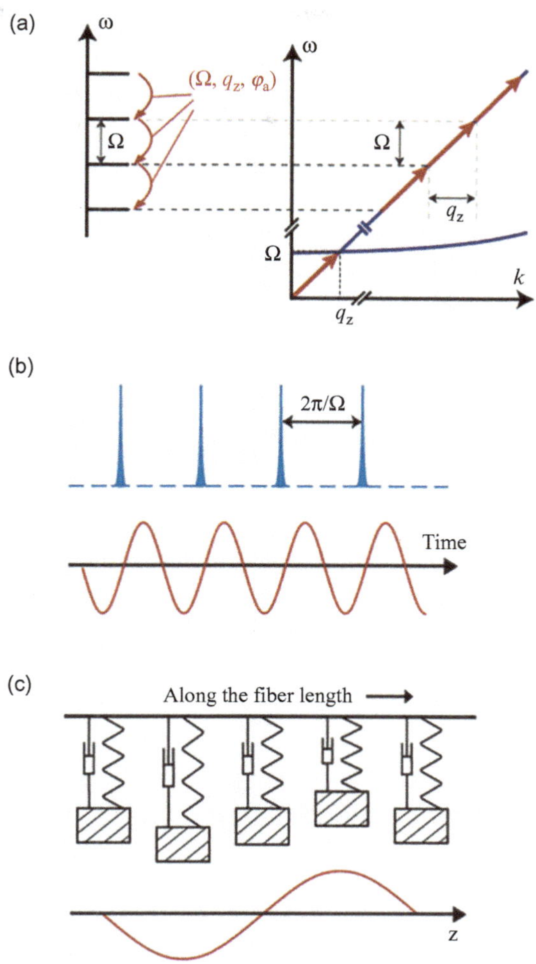

Fig. 5 (A) Dispersion diagrams of forward Brillouin scattering exhibiting cascaded effects; (B) and (C) Conceptual illustration of interactions between an mechanical wave and a train of optical pulses in cascaded forward Brillouin scattering processes.

scattering are almost transverse when close to the cut-off. In fact, considering the different frequencies of the optical waves, one can show that the propagation constant of the mechanical wave along the fiber is $q_z \approx \Omega/(c_{vac}/n_{eff})$, where Ω is the mechanical frequency, c_{vac} is the speed of light in vacuum and n_{eff} is the effective index of the optical mode. Considering a 2 GHz mode, this gives a very long spatial period along the fiber length, on the order of ten centimeters. One can then picture the mechanical mode as a row of localized, transverse mechanical oscillators (see Fig. 5C). Notice that the phase velocity of the mechanical wave near the cut-off is very high, in fact, it is equal to the velocity of optical wave $\Omega/q_z \approx c_{vac}/n_{eff}$.

When the interactions between the transverse mechanical wave and the optical pulse train in solid-core PCF are introduced in a pulsed fiber laser cavity, a few interesting phenomena can be observed. This is because in such an optomechanical laser system, the back-action of the mechanical wave on the optical pulses not only modulates the carrier frequency of the pulses, but also changes the velocity and therefore the relative positions of the pulses in the cavity due to group velocity dispersion (Kang et al., 2013; Pang et al., 2015, 2016). A conceptual illustration of an optomechanical laser cavity, with one mirror mounted on a mechanical resonator, is shown in Fig. 6A. A train of optical pulses with a repetition rate lying within the optomechanical gain bandwidth coherently drives this mechanical oscillator through electrostriction or radiation pressure, while the optically driven mechanical oscillation modulates the cavity length through elasto-optical or moving-boundary effects, providing optomechanical feedback. As a result, each pulse is locked within a different vibration cycle (see Fig. 6B), creating optomechanically bound pulse states (He et al., 2019; Pang et al., 2016).

The pulse train in the optomechanical cavity can be viewed as a set of interleaved "sublasers," each of which is mode-locked at the fundamental cavity round-trip frequency. The idea of using optomechanical interactions to mode-lock fiber lasers was firstly demonstrated in an Er-doped fiber laser working in the soliton (anomalous dispersion) regime (Kang et al., 2013; Pang et al., 2015), and then the duration of pulses was considerably

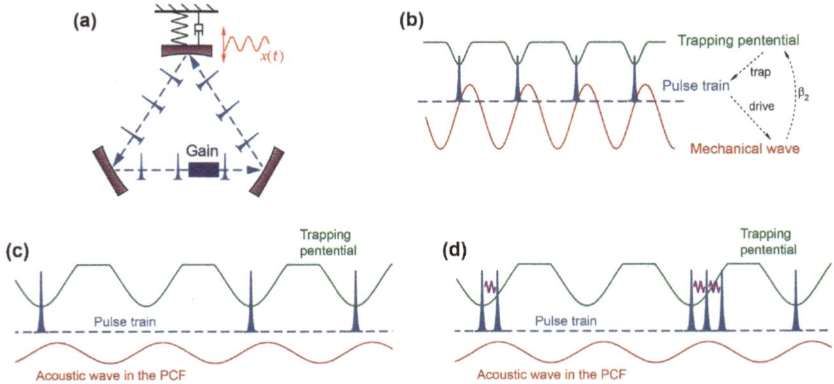

Fig. 6 (A) Pulsed laser cavity coupled with a mechanical resonator; (B) Strong interactions between pulsed light and a mechanical wave result in the stable locking state; (C) and (D) Both "empty" and "multipulse" cycles can also be stable.

shortened using careful intracavity dispersion management (He et al., 2019), leading to the generation of GHz-rate pulse trains with sub-100 fs pulse duration. High-harmonic mode-locking, in which each cycle of the mechanical vibration is occupied by one optical pulse (see Fig. 6C), is merely one example of stable states that can be obtained in the system. In practice, pulse-sequence states leaving some "empty" vibration cycles or some cycles with "multiple pulses" trapped inside can also be stable (see Fig. 6D), provided that locking between the optical and the mechanical wave is maintained. This permits encoded pulse sequences at GHz repetition rate to be generated and stored long-term (He et al., 2019; Pang et al., 2016). Recently, some interesting phenomena concerning soliton-molecule dynamics and cooscillations of multiple mechanical resonances were also demonstrated in such optomechanical pulsed-laser systems (He et al., 2021; Yeh et al., 2021).

3. Optomechanical interaction in hollow-core PCFs

We now discuss the properties of Brillouin scattering in hollow-core photonic crystal fibers (HC-PCFs). As mentioned in the introduction, these fibers guide light mostly in the central hollow-core with relatively low attenuation, enabling nonlinear interaction with gases or liquids for relatively long distances. In addition, controlling the gas pressure enables tuning the nonlinear coefficient as well as the overall waveguide dispersion, and therefore represents an attractive mechanism to control nonlinear interactions such as Kerr, Raman, and Brillouin (Benabid et al., 2002b; Nold et al., 2010; Yang et al., 2020).

3.1 Forward and backward Brillouin scattering

Forward Brillouin scattering in hollow-core fibers can arise from the interaction of light and mechanical modes of different origins (Iyer et al., 2020; Renninger et al., 2016a,b; Zhong et al., 2015). First, as illustrated in Fig. 1L, mechanical modes supported by the surrounding silica microstructured cladding induce surprisingly strong optomechanical coupling. Despite relatively low overlap with the optical fields, these modes induce strong moving-boundary effect—in contrast with conventional solid-core fibers, where elastooptic coupling is orders of magnitude larger (Renninger et al., 2016b). The second kind of modes are pressure waves in the central air region, as illustrated in Fig. 1K. Compared with the cladding modes, these pressure waves have much stronger overlap with the optical modes, as both are concentrated within the fiber core. Due to weak photoelastic response and strong damping, the overall Brillouin gain arising from these air-modes is comparable with that from silica cladding mechanical modes (Renninger et al., 2016a).

On the other hand, *stimulated* backward scattering can lead to large optical amplification in a gas-filled HC-PCF, as demonstrated in Yang et al. (2020). The interaction involves again pressure waves in the central region, filled with carbon dioxide (CO_2) at a pressure of 41 bar (see Fig. 1M). A large gain of 0.32 dB/mW over only a 50 m long fiber was observed, resulting primarily from two factors. First, as already mentioned both optical and pressure waves are tightly confined in the core and can interact for long lengths. Second, and perhaps more interestingly, the peak Brillouin gain coefficient depends approximately *quadratically* on the gas pressure, in contrast with stimulated Raman scattering (Benabid et al., 2002b) and Kerr nonlinearity (Russell et al., 2014) in which the nonlinear coefficients are linearly proportional to the gas pressure (Travers et al., 2011). This effect is further discussed in the next section.

3.2 Brillouin gain and dissipation in gas-filled fibers

During the SBS interaction, the probe wave is amplified with a peak gain given by Kobyakov et al. (2010):

$$g_B = \frac{\gamma_e^2 \omega^2}{\rho n v_a c^3 \Gamma_B A_{eff}^{ao}}, \tag{1}$$

where γ_e is the electrostrictive constant in the gas medium, ω is the light angular frequency, ρ is the gas density, n is the gas refractive index, c is the light velocity in vacuum. Here, $\Gamma_B/2\pi$ is the gain spectrum linewidth (full width at half maximum, FWHM), and is directly related to the mechanical attenuation. Fig. 7 shows the spatial distributions of the fundamental optical and first excited radial mechanical eigenmodes, respectively (obtained by a finite element method using COMSOL Multiphysics). The acousto-optic overlap effective area (A_{eff}^{ao}) is calculated from the overlap integral between these two distributions (Kobyakov et al., 2010).

We now discuss the dependencies of the key parameters in Eq. (1) leading to very large gain. First, the mechanical velocity is given by $v_a = 1/\sqrt{\beta_s \cdot \rho}$, independent of pressure for an ideal gas, where β_s is the adiabatic compressibility which is inversely proportional to the gas pressure in ideal gas model. As the compressibility coefficient of gases (e.g., CO_2 gas at 40 bar) is four orders

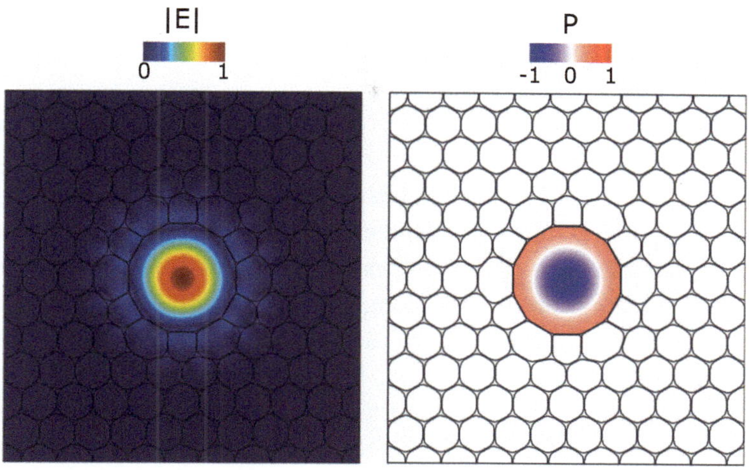

Fig. 7 (A) Normalized optical intensity distribution of the fundamental optical mode; (B) Normalized pressure distribution of the first excited radial acoustic mode.

of magnitude larger than that of solid materials (e.g., fused silica) and the density of gases (at 40 bar) is some 30 times smaller than that of a condensed material, the mechanical velocity is about 20 times smaller than in silica. This leads to a 20-fold increase in gain, as shown by Eq. (1). It also leads to a Brillouin frequency shift ν_B lying in the subgigahertz range (e.g., ~320 MHz for CO_2 at a pump wavelength of 1.55 μm).

The electrostrictive constant, γ_e, and the mechanical attenuation coefficient, Γ_B, are the two key parameters resulting in the quadratic dependence of the Brillouin gain coefficient on gas pressure. The electrostrictive constant is defined as the normalized rate of change of the relative permittivity ϵ_r in response to a change in the density ρ (Boyd, 2008): $\gamma_e = \rho \frac{\partial \epsilon_r}{\partial \rho} = \rho \frac{\partial \chi}{\partial \rho}$, where χ is the electric susceptibility. Normally, the electric susceptibility of gas shows a linear dependence on density and as a result, the electrostrictive constant is directly proportional to the density, hence to the pressure in the ideal gas approximation. The mechanical attenuation coefficient can be represented by Landau and Lifshitz (1987): $\Gamma_B = \frac{q^2}{\rho}\left[\frac{4}{3}\eta_s + \eta_b + \frac{\kappa}{C_P}(\gamma - 1)\right]$, where η_s and η_b are the shear and bulk viscosity respectively, κ is the thermal conductivity and C_P is the specific heat at constant pressure. Therefore, the mechanical attenuation coefficient is inversely proportional to the density of the gas and hence to the gas pressure in the ideal gas approximation. Inserting both the electrostrictive constant and the mechanical attenuation coefficient into

Eq. (1), we conclude that the Brillouin gain coefficient increases quadratically with the density, and so with the pressure.

3.3 Applications

The high Brillouin coefficient in gas-filled HC-PCF can be exploited for various novel optical functionalities and devices such as in sensing and lasing (Fig. 8). One example is for *strain-insensitive* distributed temperature sensing (Fig. 8A), which addresses the long-standing strain temperature cross-sensitivity problem in previous Brillouin fiber sensors. The optical mode concentrated in the hollow-core is almost insensitive to a strain applied on the solid cladding. In solid-core fibers, on the other hand, both strain and temperature variation unavoidably change the refractive index (and acoustic velocity) and hence the Brillouin frequency shift. In Yang et al. (2020), distributed temperature sensing system without strain cross-sensitivity

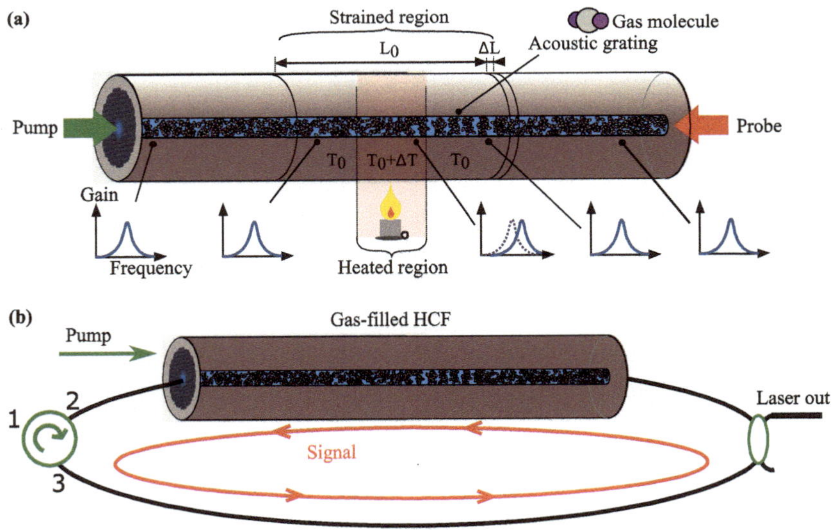

Fig. 8 Sensing and lasing applications based on gas-filled HC-PCF (Yang et al., 2020). (A) Strain-insensitive distributed temperature sensing in gas-filled HC-PCF using stimulated Brillouin scattering. An acoustic grating is generated in the gas-filled HC-PCF by pump and probe interaction. Gas is not responsive to any strain applied to the fiber. However, the acoustic velocity is significantly changed when a temperature variation is present, resulting in a frequency shift for the Brillouin gain spectrum. (B) Low-threshold gas Brillouin fiber laser based on Brillouin scattering in HC-PCF. Pump is launched into the gas-filled HC-PCF through a circulator. The gain coefficient and linewidth can be tuned by changing the gas pressure in the HC-PCF and using different type of gas.

has been experimentally demonstrated. It must be mentioned that since the Brillouin gain coefficient in 41 bar CO_2 filled HC-PCF is 6 times higher than the standard single-mode fiber and the Brillouin linewidth in HC-PCF is 10 times narrower than the SMF, improved sensing performance is observed. In fact, ~1 cm spatial resolution, 0.3 degree temperature resolution and 1.2 MHz/°C temperature response have been achieved in HC-PCF.

As another application, a gas Brillouin laser can be generated by configuring a gas-filled HC-PCF as a fiber ring cavity (Fig. 8B). In Yang et al. (2020), a single-pass pump configuration with a 41 bar CO_2 filled HC-PCF has been used and the gas Brillouin laser was turned on when the net Brillouin gain of the Stokes light exceeds the round-trip loss of the fiber cavity. The measured lasing threshold was 33.2 mW, which was in good agreement with the theoretical estimated threshold of 35 mW. It has also been demonstrated that the laser linewidth (full width half maximum) was measured to be 66 kHz which is much narrower than the spontaneous spectrum of the Brillouin scattering in a 41 bar CO_2 filled HC-PCF (3.66 MHz). The gas Brillouin laser in HC-PCF provides also improved tunability compared to that in solid-core fiber because the Brillouin gain and the gain linewidth can be changed by using different gas pressure as well as different type of gas.

4. Conclusions

In the chapter, we discussed the properties of optical and mechanical modes supported by PCFs, how these properties impact the optomechanical interaction and highlighted several applications that have leveraged their unique properties. We discussed how confinement of optical and mechanical waves in the core region together with long interaction lengths makes this an interesting platform, as both conditions are difficult to simultaneously achieve in other waveguides such as conventional solid fibers or integrated waveguides.

An overview of the nature of mechanical modes supported by this fibers was provided, and how they lead to variety of interesting phenomena. For example, tightly confined axial–radial modes lead to single-frequency forward scattering resembling Raman scattering, and has been leveraged high repetition rate mode-locking fiber lasers and to explore storage architectures. Torsional-radial modes inducing polarization coupling in a highly birefringent fiber were leveraged to stimulated forward scattering and was the basis to demonstrate an efficient optical isolator. Surface waves concentrated in the air-glass interface of solid-core fibers have been observed and flexural waves in which the core itself oscillates almost like a free-standing rod have been leveraged to study intermodal scattering.

We have also demonstrated how the air-glass boundaries enhances the effect of moving-boundary effect, in some cases reinforcing and in others suppressing the effect of photoelasticity, depending on the specific profiles of the mechanical and optical modes that interact. Finally, we reviewed Brillouin scattering arising from acoustic waves in the high pressure gas cavity in hollow-core, which were demonstrated to induce strong stimulated backward scattering. These results, accumulated over more than a decade of research on Brillouin scattering in PCFs, have established this as an important platform to study fundamental aspects of optomechanical interaction in waveguides, and can be advantageously used to enhance and harness Brillouin scattering in view of potential applications to fiber lasers, fiber-optic sensors and all-optical processing.

References

Auld, B.A., Green, R.E., 1974. Acoustic fields and waves in solids: two volumes. Phys. Today 27 (10), 63.

Benabid, F., Knight, J.C., Antonopoulos, G., Russell, P.S.J., 2002a. Stimulated Raman scattering in hydrogen-filled hollow-core photonic crystal fiber. Science 298 (5592), 399–402.

Benabid, F., Knight, J.C., Antonopoulos, G., Russell, P.S.J., 2002b. Stimulated Raman scattering in hydrogen-filled hollow-core photonic crystal fiber. Science 298 (5592), 399–402. https://doi.org/10.1126/science.1076408.

Beugnot, J.-C., Laude, V., 2012. Electrostriction and guidance of acoustic phonons in optical fibers. Phys. Rev. B 86 (22), 224304.

Beugnot, J.-C., Sylvestre, T., Alasia, D., Maillotte, H., Laude, V., Monteville, A., Provino, L., Traynor, N., Mafang, S.F., Thévenaz, L., 2007a. Complete experimental characterization of stimulated Brillouin scattering in photonic crystal fiber. Opt. Express 15 (23), 15517–15522.

Beugnot, J.-C., Sylvestre, T., Maillotte, H., Mélin, G., Laude, V., 2007b. Guided acoustic wave Brillouin scattering in photonic crystal fibers. Opt. Lett. 32 (1), 17–19.

Beugnot, J.-C., Lebrun, S., Pauliat, G., Maillotte, H., Laude, V., Sylvestre, T., 2014. Brillouin light scattering from surface acoustic waves in a subwavelength-diameter optical fibre. Nat. Commun. 5 (1), 1–6.

Boyd, R.W., 2008. Nonlinear Optics, third ed. Academic Press. https://www.elsevier.com/books/nonlinear-optics/boyd/978-0-12-369470-6.

Brenn, A., Wiederhecker, G.S., Kang, M.S., Hundertmark, H., Joly, N., Russell, P.S.J., 2009. Influence of air-filling fraction on forward Raman-like scattering by transversely trapped acoustic resonances in photonic crystal fibers. J. Opt. Soc. Am. B 26 (8), 1641–1648.

Broderick, N.G.R., Monro, T.M., Bennett, P.J., Richardson, D.J., 1999. Nonlinearity in holey optical fibers: measurement and future opportunities. Opt. Lett. 24 (20), 1395–1397.

Carry, E., Beugnot, J.-C., Stiller, B., Lee, M.W., Maillotte, H., Sylvestre, T., 2011. Temperature coefficient of the high-frequency guided acoustic mode in a photonic crystal fiber. Appl. Opt. 50 (35), 6543–6547.

Cheng, T., Cherif, R., Liao, M., Gao, W., Duan, Z., Zghal, M., Suzuki, T., Ohishi, Y., 2012. Stimulated Brillouin scattering of higher-order acoustic modes in four-core tellurite microstructured optical fiber. Appl. Phys. Express 5 (10), 102501.

Cherif, R., Zghal, M., Tartara, L., 2012. Characterization of stimulated Brillouin scattering in small core microstructured chalcogenide fiber. Opt. Commun. 285 (3), 341–346.

Dainese, P., Russell, P.S.J., Joly, N., Knight, J.C., Wiederhecker, G.S., Fragnito, H., Laude, V., Khelif, A., 2006a. Stimulated Brillouin scattering from multi-GHz-guided acoustic phonons in nanostructured photonic crystal fibres. Nat. Phys. 2 (6), 388–392.

Dainese, P., Russell, P.S.J., Wiederhecker, G.S., Joly, N., Fragnito, H.L., Laude, V., Khelif, A., 2006b. Raman-like light scattering from acoustic phonons in photonic crystal fiber. Opt. Express 14 (9), 4141–4150.

De Matos, C.J.S., Taylor, J.R., Hansen, K.P., 2004. All-fibre Brillouin laser based on holey fibre yielding comb-like spectra. Opt. Commun. 238 (1–3), 185–189.

Dudley, J.M., Genty, G., Coen, S., 2006. Supercontinuum generation in photonic crystal fiber. Rev. Mod. Phys. 78 (4), 1135.

Elser, D., Andersen, U.L., Korn, A., Glöckl, O., Lorenz, S., Marquardt, C., Leuchs, G., 2006. Reduction of guided acoustic wave Brillouin scattering in photonic crystal fibers. Phys. Rev. Lett. 97 (13), 133901.

Fatome, J., Fortier, C., Nguyen, T.N., Chartier, T., Smektala, F., Messaad, K., Kibler, B., Pitois, S., Gadret, G., Finot, C., et al., 2009. Linear and nonlinear characterizations of chalcogenide photonic crystal fibers. J. Lightwave Technol. 27 (11), 1707–1715.

Florez, O., Jarschel, P., Espinel, Y.A.V., Cordeiro, C.M.B., Alegre, T.P., Wiederhecker, G., Dainese, P., 2016. Brillouin scattering self-cancellation. Nat. Commun. 7 (1), 1–8.

Fortier, C., Fatome, J., Pitois, S., Smektala, F., Millot, G., Troles, J., Desevedavy, F., Houizot, P., Brilland, L., Traynor, N., 2008. Experimental investigation of Brillouin and Raman scattering in a 2SG sulfide glass microstructured chalcogenide fiber. Opt. Express 16 (13), 9398–9404.

Furusawa, K., Yusoff, Z., Poletti, F., Monro, T.M., Broderick, N.G.R., Richardson, D.J., 2006. Brillouin characterization of holey optical fibers. Opt. Lett. 31 (17), 2541–2543.

Harun, S.W., Aziz, S.N., Tamchek, N., Shahabuddin, N.S., Ahmad, H., 2008. Brillouin fibre laser with 20 m-long photonic crystal fibre. Electron. Lett. 44 (18), 1065–1066.

He, W., Pang, M., Yeh, D.-H., Huang, J., Menyuk, C.R., Russell, P.S.J., 2019. Formation of optical supramolecular structures in a fibre laser by tailoring long-range soliton interactions. Nat. Commun. 10 (1), 1–9.

He, W., Pang, M., Yeh, D.-H., Huang, J., Russell, P.S.J., 2021. Synthesis and dissociation of soliton molecules in parallel optical-soliton reactors. Light Sci. Appl. 10, 120.

Hong, K.S., Lee, K.J., Yeom, D.-I., Kim, B.Y., 2010. Characterization of structural irregularities in highly birefringent photonic crystal fiber using torsional acoustic polarization coupling. Opt. Commun. 283 (20), 4094–4098.

Iyer, A., Xu, W., Antonio-Lopez, J.E., Correa, R.A., Renninger, W.H., 2020. Ultra-low Brillouin scattering in anti-resonant hollow-core fibers. APL Photonics 5 (9), 096109. https://doi.org/10.1063/5.0017796.

Jarschel, P.F., Magalhaes, L.S., Aldaya, I., Florez, O., Dainese, P., 2018. Fiber taper diameter characterization using forward Brillouin scattering. Opt. Lett. 43 (5), 995–998.

Jarschel, P.F., Lamilla, E., Espinel, Y.A.V., Aldaya, I., Pita, J.L., Gil-Molina, A., Wiederhecker, G.S., Dainese, P., 2021. Intermodal Brillouin scattering in solid-core photonic crystal fibers. APL Photonics 6 (3), 036108.

Kang, M.S., Brenn, A., Wiederhecker, G.S., Russell, P.S.J., 2008. Optical excitation and characterization of gigahertz acoustic resonances in optical fiber tapers. Appl. Phys. Lett. 93 (13), 131110.

Kang, M.S., Nazarkin, A., Brenn, A., Russell, P., 2009. Tightly trapped acoustic phonons in photonic crystal fibres as highly nonlinear artificial Raman oscillators. Nat. Phys. 5 (4), 276–280.

Kang, M.S., Brenn, A., Russell, P.S.J., 2010. All-optical control of gigahertz acoustic resonances by forward stimulated interpolarization scattering in a photonic crystal fiber. Phys. Rev. Lett. 105 (15), 153901.

Kang, M.S., Butsch, A., Russell, P.S.J., 2011. Reconfigurable light-driven opto-acoustic isolators in photonic crystal fibre. Nat. Photonics 5 (9), 549–553.

Kang, M.S., Joly, N.Y., Russell, P.S.J., 2013. Passive mode-locking of fiber ring laser at the 337th harmonic using gigahertz acoustic core resonances. Opt. Lett. 38 (4), 561–563.

Kobyakov, A., Sauer, M., Chowdhury, D., 2010. Stimulated Brillouin scattering in optical fibers. Adv. Opt. Photon. 2 (1), 1–59. https://doi.org/10.1364/AOP.2.000001.

Landau, L.D., Lifshitz, E.M., 1987. Fluid Mechanics, second ed. Pergamon Press, Pergamon, Oxford.

Laude, V., Khelif, A., Benchabane, S., Wilm, M., Sylvestre, T., Kibler, B., Mussot, A., Dudley, J.M., Maillotte, H., 2005. Phononic band-gap guidance of acoustic modes in photonic crystal fibers. Phys. Rev. B 71 (4), 045107.

Lee, J.H., Yusoff, Z., Belardi, W., Ibsen, M., Monro, T.M., Richardson, D.J., 2002. Investigation of Brillouin effects in small-core holey optical fiber: lasing and scattering. Opt. Lett. 27 (11), 927–929.

Liang, H., Li, J., Han, B., Chang, Y., Cheng, L., Guan, B.-O., 2013. Potential for simultaneous strain and temperature sensing based on Brillouin scattering in an all-solid photonic bandgap fiber. Opt. Lett. 38 (4), 465–467.

Limpert, J., Schreiber, T., Nolte, S., Zellmer, H., Tuennermann, A., Iliew, R., Lederer, F., Broeng, J., Vienne, G., Petersson, A., et al., 2003. High-power air-clad large-mode-area photonic crystal fiber laser. Opt. Express 11 (7), 818–823.

McElhenny, J.E., Pattnaik, R.K., Toulouse, J., Saitoh, K., Koshiba, M., 2008. Unique characteristic features of stimulated Brillouin scattering in small-core photonic crystal fibers. J. Opt. Soc. Am. B 25 (4), 582–593.

Minardo, A., Bernini, R., Urbanczyk, W., Wojcik, J., Gorbatov, N., Tur, M., Zeni, L., 2008. Stimulated Brillouin scattering in highly birefringent microstructure fiber: experimental analysis. Opt. Lett. 33 (20), 2329–2331.

Nasir, M.N.M., Yusoff, Z., Al-Mansoori, M.H., Rashid, H.A.A., Choudhury, P.K., 2009. Widely tunable multi-wavelength Brillouin-erbium fiber laser utilizing low SBS threshold photonic crystal fiber. Opt. Express 17 (15), 12829–12834.

Nold, J., Hölzer, P., Joly, N., Wong, G.K.L., Nazarkin, A., Podlipensky, A., Scharrer, M., Russell, P., 2010. Pressure-controlled phase matching to third harmonic in Ar-filled hollow-core photonic crystal fiber. Opt. Lett. 35 (17), 2922–2924.

Pang, M., Jiang, X., He, W., Wong, G.K.L., Onishchukov, G., Joly, N.Y., Ahmed, G., Menyuk, C.R., Russell, P.S.J., 2015. Stable subpicosecond soliton fiber laser passively mode-locked by gigahertz acoustic resonance in photonic crystal fiber core. Optica 2 (4), 339–342.

Pang, M., He, W., Jiang, X., Russell, P.S.J., 2016. All-optical bit storage in a fibre laser by optomechanically bound states of solitons. Nat. Photonics 10 (7), 454–458.

Poletti, F., Furusawa, K., Yusoff, Z., Broderick, N.G.R., Richardson, D.J., 2007. Nonlinear tapered holey fibers with high stimulated Brillouin scattering threshold and controlled dispersion. J. Opt. Soc. Am. B 24 (9), 2185–2194.

Renninger, W.H., Behunin, R.O., Rakich, P.T., 2016a. Guided-wave Brillouin scattering in air. Optica 3 (12), 1316–1319.

Renninger, W.H., Shin, H., Behunin, R.O., Kharel, P., Kittlaus, E.A., Rakich, P.T., 2016b. Forward Brillouin scattering in hollow-core photonic bandgap fibers. New J. Phys. 18 (2), 025008.

Russell, P., 2003. Photonic crystal fibers. Science 299 (5605), 358–362.

Russell, P.S.J., Marin, E., Diez, A., Guenneau, S., Movchan, A.B., 2003. Sonic band gaps in PCF preforms: enhancing the interaction of sound and light. Opt. Express 11 (20), 2555–2560.

Russell, P.S.J., Hölzer, P., Chang, W., Abdolvand, A., Travers, J.C., 2014. Hollow-core photonic crystal fibres for gas-based nonlinear optics. Nat. Photonics 8 (4), 278–286.

Shahi, S., Harun, S.W., Ahmad, H., 2009. Multi-wavelength Brillouin fiber laser using a holey fiber and a bismuth-oxide based erbium-doped fiber. Laser Phys. Lett. 6 (6), 454.

Shelby, R.M., Levenson, M.D., Bayer, P.W., 1985. Guided acoustic-wave Brillouin scattering. Phys. Rev. B 0163-1829. 31 (8), 5244–5252. https://doi.org/10.1103/PhysRevB.31.5244.

Shibata, N., Nakazono, A., Taguchi, N., Tanaka, S., 2006. Forward Brillouin scattering in holey fibers. IEEE Photon. Technol. Lett. 18 (2), 412–414.

Skryabin, D.V., Luan, F., Knight, J.C., Russell, P.S.J., 2003. Soliton self-frequency shift cancellation in photonic crystal fibers. Science 301 (5640), 1705–1708.

Smith, R.G., 1972. Optical power handling capacity of low loss optical fibers as determined by stimulated Raman and Brillouin scattering. Appl. Opt. 11 (11), 2489–2494.

Soltanian, M.R.K., Ahmad, H., Khodaie, A., Amiri, I.S., Ismail, M.F., Harun, S.W., 2015. A stable dual-wavelength thulium-doped fiber laser at 1.9 μm using photonic crystal fiber. Sci. Rep. 5 (1), 1–8.

Stiller, B., Foaleng, S.M., Beugnot, J.-C., Lee, M.W., Delqué, M., Bouwmans, G., Kudlinski, A., Thévenaz, L., Maillotte, H., Sylvestre, T., 2010. Photonic crystal fiber mapping using Brillouin echoes distributed sensing. Opt. Express 18 (19), 20136–20142.

Stiller, B., Delqué, M., Beugnot, J.-C., Lee, M.W., Mélin, G., Maillotte, H., Laude, V., Sylvestre, T., 2011. Frequency-selective excitation of guided acoustic modes in a photonic crystal fiber. Opt. Express 19 (8), 7689–7694.

Stiller, B., Kudlinski, A., Lee, M.W., Bouwmans, G., Delque, M., Beugnot, J.-C., Maillotte, H., Sylvestre, T., 2012. SBS mitigation in a microstructured optical fiber by periodically varying the core diameter. IEEE Photon. Technol. Lett. 24 (8), 667–669.

Tchahame, J.C., Beugnot, J.-C., Kudlinski, A., Sylvestre, T., 2015. Multimode Brillouin spectrum in a long tapered birefringent photonic crystal fiber. Opt. Lett. 40 (18), 4281–4284. https://doi.org/10.1364/OL.40.004281.

Tchahame, J.C., Beugnot, J.-C., Huy, K.P., Laude, V., Kudlinski, A., Sylvestre, T., 2016. Surface Brillouin scattering in photonic crystal fibers. Opt. Lett. 41 (14), 3269–3272.

Thurston, R.N., 1978. Elastic waves in rods and clad rods. J. Acoust. Soc. Am. 64 (1), 1–37. https://doi.org/10.1121/1.381962.

Travers, J.C., Chang, W., Nold, J., Joly, N.Y., Russell, P.S.J., 2011. Ultrafast nonlinear optics in gas-filled hollow-core photonic crystal fibers. J. Opt. Soc. Am. B 1520-8540. 28 (12), A11–A26. https://doi.org/10.1364/JOSAB.28.000A11.

Wadsworth, W.J., Joly, N., Knight, J.C., Birks, T.A., Biancalana, F., Russell, P.S.J., 2004. Supercontinuum and four-wave mixing with Q-switched pulses in endlessly single-mode photonic crystal fibres. Opt. Express 12 (2), 299–309.

Waldron, R.A., 1969. Some problems in theory of guided microsonic waves. IEEE Trans. Microwave Theory Tech. 17 (11), 893–904. http://ieeexplore.ieee.org/lpdocs/epic03/wrapper.htm?arnumber=1127079 papers3://publication/doi/10.1109/Tmtt.1969.1127079.

Wiederhecker, G.S., Brenn, A., Fragnito, H.L., Russell, P.S.J., 2008. Coherent control of ultrahigh-frequency acoustic resonances in photonic crystal fibers. Phys. Rev. Lett. 100 (20), 203903.

Wiederhecker, G.S., Dainese, P., Mayer Alegre, T.P., 2019. Brillouin optomechanics in nanophotonic structures. APL Photonics 4 (7), 071101.

Woodward, R.I., Kelleher, E.J.R., Popov, S.V., Taylor, J.R., 2014. Stimulated Brillouin scattering of visible light in small-core photonic crystal fibers. Opt. Lett. 39 (8), 2330–2333.

Yang, S., Chen, H., Qiu, C., Chen, M., Chen, M., Xie, S., Li, J., Chen, W., 2008. Slow-light delay enhancement in small-core pure silica photonic crystal fiber based on Brillouin scattering. Opt. Lett. 33 (2), 95–97.

Yang, F., Gyger, F., Thévenaz, L., 2020. Intense Brillouin amplification in gas using hollow-core waveguides. Nat. Photonics 14 (11), 700–708.

Yeh, D.-H., He, W., Pang, M., Jiang, X., Russell, P.S.J., 2021. Synchronization of gigahertz core resonances in multiple photonic crystal fiber cores by timing-modulated harmonic mode locking. Optica 8 (12), 1581–1585.

Zeng, X., He, W., Frosz, M.H., Geilen, A., Roth, P., Wong, G.K.L., Russell, P.S.J., Stiller, B., 2021. Stimulated Brillouin scattering in chiral photonic crystal fiber. arXiv preprint arXiv:2111.05289.

Zhong, W.E., Stiller, B., Elser, D., Heim, B., Marquardt, C., Leuchs, G., 2015. Depolarized guided acoustic wave Brillouin scattering in hollow-core photonic crystal fibers. Opt. Express 23 (21), 27707–27714.

Zou, L., Bao, X., Chen, L., 2003. Brillouin scattering spectrum in photonic crystal fiber with a partially germanium-doped core. Opt. Lett. 28 (21), 2022–2024.

Zou, L., Bao, X., Afshar, S., Chen, L., 2004. Dependence of the Brillouin frequency shift on strain and temperature in a photonic crystal fiber. Opt. Lett. 29 (13), 1485–1487.

Zou, W., He, Z., Hotate, K., 2012. Experimental investigation on Brillouin scattering property in highly nonlinear photonic crystal fiber with hybrid core. Opt. Express 20 (10), 11083–11090.

CHAPTER FIVE

On-chip stimulated Brillouin scattering

Ravi Pant[a,*], Stephen J. Madden[b], Benjamin J. Eggleton[c], and Heedeuk Shin[d]

[a]School of Physics, Indian Institute of Science Education and Research (IISER), Thiruvananthapuram, Kerala, India
[b]Laser Physics Center, Australian National University, Canberra, ACT, Australia
[c]School of Physics, University of Sydney Nano Institute (Sydney Nano), University of Sydney, NSW, Australia
[d]Department of Physics, Pohang University of Science and Technology (POSTECH), Pohang, South Korea
*Corresponding author: e-mail address: rpant@iisertvm.ac.in

Contents

In the preceding chapters, we have seen how the development of optical fibers reduced the threshold power for the excitation of Brillouin scattering to a few mW from tens-of-Watts in bulk media (Ippen and Stolen, 1972; Kobyakov et al., 2010). Observation of Brillouin scattering was, therefore, possible with low-power semiconductor lasers. The long interaction length and large optical intensity, which results from the small mode area in a fiber, caused this drastic reduction in the threshold power. Many real-world applications, however, require miniaturized, robust, low-cost devices where potentially the whole system is integrated on a single chip. Excitation of stimulated Brillouin scattering (SBS) in a chip-level device is, therefore, important to harness SBS for real-world applications. In this chapter, we discuss on-chip stimulated Brillouin scattering and its applications. Section 1 discusses the requirements for realizing SBS in a chip-scale device.

Semiconductors and Semimetals, Volume 109
ISSN 0080-8784
https://doi.org/10.1016/bs.semsem.2022.04.003

Section 2 presents the fabrication of photonic chips made from soft glass materials, e.g., chalcogenide. Demonstration of on-chip SBS in soft glass photonic circuits is discussed in Section 3, followed by its applications to slow-light, Brillouin dynamic gratings, and mode-locked pulse generation in Section 4. In Section 5, a brief overview of hybrid integration with CMOS technology is given.

1. Material and device requirements

From the theory of SBS in chapter "Theoretical formalisms for stimulated Brillouin scattering" by Wolff et al., we have learned that a probe field E_s (E_{as}) centered at the Stokes (anti-Stokes) frequency experiences an exponential gain (loss) under the continuous-wave and undepleted pump approximation. The Stokes intensity $I_s(L)$ at the output of an SBS medium of length L is related to the input intensity $I_s(0)$ according to $I_s(L) = I_s(0)e^G$, where the gain exponent G is related to the Brillouin scattering cross-section g_B, effective interaction length L_{eff}, pump power P_p, and effective mode area A_{eff} according to $G = \frac{g_B P_p L_{eff}}{A_{eff}}$. One can, therefore, achieve large gain with small pump power in a short length device using a material with large g_B and by designing a waveguide with small A_{eff}. The Brillouin scattering cross-section $g_B \propto \eta n^7 p_{eo}^2$, where η is the acousto-optic overlap, p_{eo} is the elasto-optic coefficient of an SBS medium, and n is the refractive index at the pump wavelength (Poulton et al., 2013). Therefore, materials with large n and p_{eo} are suitable for developing on-chip SBS devices. Large index materials also allow small A_{eff} due to the availability of large index contrast between the core and cladding materials. However, having a material with large n and p_{eo} is not sufficient to achieve high g_B, strong confinement of both the optical and acoustic modes within the waveguide is necessary to have a large value of η, which varies between 0 and 1. Exciting SBS in a chip-scale device, therefore, requires the development of high-index material-based photonic structures that allow a strong overlap of both the optical and acoustic modes to achieve large g_B and small A_{eff}. However, A_{eff} of photonic structures cannot be arbitrarily reduced due to an increase in the sidewall scattering induces losses for smaller dimensions, which creates a trade-off between A_{eff} and L_{eff}. The propagation loss due to sidewall scattering is proportional to d^{-4}, where d is the waveguide width (Lee et al., 2000). Therefore, L_{eff} would increase significantly with an increase in the waveguide dimension. A device with a larger A_{eff} can therefore achieve larger gain by compensating the reduction in the pump intensity with increased length (Kim and Shin, 2019). With the recent advances in nanofabrication technology, it is possible to fabricate tens-of-cm long waveguides in a small form factor with significantly small A_{eff} and low losses (Hwang et al., 2021; Lee et al., 2012). The waveguide, however, must allow strong overlap between the optical and acoustic modes by confining them in the waveguide.

Simultaneous confinement of both the optical and acoustic modes occurs under the following conditions: (i) when both the optical and acoustic modes are guided or (ii) the optical mode is guided and acoustic mode is leaky (Poulton et al., 2013). For an optical (acoustic) mode with speeds u_1 (v_1) and u_2 (v_2) in the core and cladding, respectively, guided propagation through total internal reflection (TIR) occurs when $u_1 < u_2$ ($v_1 < v_2$). In (Poulton et al., 2013), the acoustic mode speed is close to the longitudinal mode speed. A leaky acoustic mode arises when $v_1 > v_2$. For a leaky mode, acoustic confinement occurs due to the impedance mismatch at the core-cladding interface. Fig. 1 shows the calculated Brillouin gain coefficient, as a function of the acoustic frequency Δf, under different conditions for the acoustic mode confinement in an optical waveguide. The properties of the optical and acoustic modes are shown in Fig. 1A. Since SBS in an optical waveguide is considered, we assume that the core refractive index n_1 is greater than the cladding refractive index n_2 for the optical mode and, therefore, it is always guided. For the acoustic mode, three different confinement regimes occur: (i) $v_1 < v_2$, acoustic guiding; (ii) $v_1 \approx v_2$, radiative regime; or (iii) $v_1 > v_2$, leaky mode, which are shown in Fig. 1B–D, respectively. For the acoustic guiding case in Fig. 1B, both optical and acoustic modes are tightly confined and the acousto-optic overlap is large, which results in a large Brillouin gain coefficient. When $v_1 \approx v_2$, the acoustic mode starts radiating out of the core, as shown by the large acoustic mode in Fig. 1C. Radiation of the acoustic mode reduces the acousto-optic interaction and shifts the acoustic mode frequency compared to that for the acoustic guiding case, which is shown as a dotted plot in Fig. 1C. The vertical dashed lines in Fig. 1B–D show the acoustic mode frequency in the bulk homogeneous medium of the waveguide core. When $v_1 > v_2$, even though there is no acoustic guiding, the acousto-optic overlap and thus Brillouin gain is large if $\tau_{leakage} > \tau_{phonon}$, where $\tau_{leakage}$ is the leakage time of the acoustic mode and τ_{phonon} is the acoustic phonon lifetime. For a cylindrical waveguide, the leakage time or lifetime of the acoustic mode in the waveguide is given according to (Poulton et al., 2013):

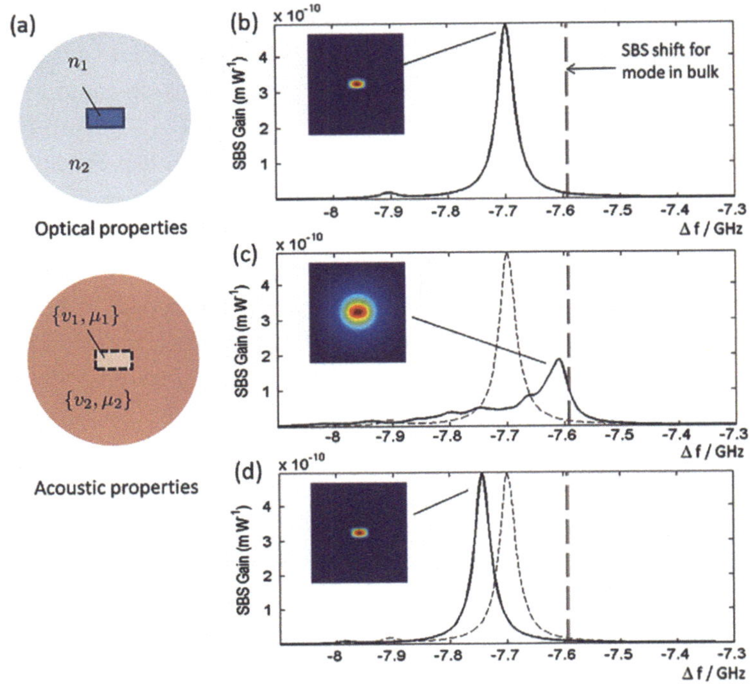

Fig. 1 Frequency dependent Brillouin gain coefficient under different acoustic confinement regimes in an optical waveguide: (A) Optical and acoustic properties of the optical waveguide device (B) Frequency dependent Brillouin gain coefficient $g_B(\Delta f)$ when the acoustic mode is guided $v_1 < v_2$. The acoustic mode is tightly confined in the optical waveguide as shown in the inset. (C) $g_B(\Delta f)$ in the radiative regime ($v_1 \approx v_2$) of the acoustic mode. The $g_B(\Delta f)$ for the acoustic guiding regime is plotted in dashed for comparison. (D) The $g_B(\Delta f)$ under the acoustic mode leakage regime $v_1 > v_2$ has nearly the same maximum value as for the acoustic guiding regime, when $\tau_{leakage} > \tau_{phonon}$, except a shift in the acoustic mode frequency compared to the guided acoustic mode (dashed curve). The vertical line in all the plots indicates the acoustic mode frequency in the bulk homogeneous medium of the core (Poulton et al., 2013).

$$\tau_{leakage} \sim \frac{\Delta f^3 a^3 \mu_2}{v_1^2 J_{00}^2 \mu_1} \sqrt{\frac{1}{v_2^2} - \frac{1}{v_1^2}}, \tag{1}$$

where a is the radius of the waveguide core, μ_1 and μ_2 are the shear modulus of the core and cladding, respectively, J_{00} is the first-zero of the Bessel function $J_0(z)$, and Δf is the acoustic mode frequency. The $\tau_{leakage}$, therefore, reduces with the waveguide dimensions. For $\tau_{leakage} < \tau_{phonon}$, the Brillouin gain profile broadens compared to the guided acoustic mode case, reducing the Brillouin gain coefficient. Having a material with a large refractive index is, therefore, not sufficient to realize a device with large g_B, and the device should confine both the acoustic and optical modes for a sufficiently long duration to have substantial interaction.

2. Device fabrication

Table 1 lists the optical and mechanical properties of some of the commonly used optical materials. Among the materials shown in the Table, silicon has the largest value of n. Silicon waveguides, therefore, strongly guide the optical mode due to their large index contrast with the surrounding media. The large index contrast between the core and cladding results in smaller optical mode area in silicon, which enhances the pump intensity. However, acoustic mode confinement in silicon is challenging as the mode rapidly ($\tau_{leakage} < \tau_{phonon}$) leaks through the silica under cladding into the silicon substrate and is lost to unguided modes. A typical silicon waveguide, therefore, does not support efficient Brillouin nonlinearity.

Chalcogenide glasses have a larger refractive index and smaller acoustic speed compared to silica (see Table 1), which makes chalcogenide waveguides with silica cladding a candidate platform for realizing SBS in a chip-scale device. The refractive indices of As_2S_3 and As_2Se_3 glasses are 1.7 and 1.9 times that of silica, respectively. The n^7 dependence of the Brillouin gain coefficient, therefore, suggests that the g_B for chalcogenide waveguides can be 40 to 100 times larger compared to silica. Large index contrast between silica and chalcogenide allows the design of chalcogenide

Table 1 Optical and acoustic properties of commonly used optical materials.

S. No.	Material	Refractive index(n)	V_a (m/s)	p_{12}	g_B (m/W)
1	Silica (SiO_2) (Abedin, 2005)	1.45	5960	0.286	4.5×10^{-11}
2	Arsenic Triselenide (As_2Se_3) (Abedin, 2005)	2.81	2250	0.266	6.0×10^{-9}
3	Arsenic Trisulphide (As_2S_3) (Eggleton et al., 2013)	2.45	–	0240	5.7×10^{-9}
4	Tellurite (Abedin, 2006)	2.03	2997		2.2×10^{-10}
5	Silicon (Si) (Eggleton et al., 2013)	3.48	9000	0.017	
6	Silicon Nitride (Si_3N_4) (Gyger et al., 2020)	2.00	1875	0.047	4.0×10^{-13}

waveguides with a much smaller optical mode area than a typical single-mode optical fiber, which has a mode area of 80 μm^2. Since the gain exponent G is proportional to g_B and varies inversely with mode area, efficient Brillouin gain is possible in cm-scale chalcogenide waveguides. The first demonstration of on-chip SBS was reported in a 7 cm long chalcogenide (As_2S_3) rib waveguide with a top silica cladding (Pant et al., 2011b). Lower acoustic and optical mode speeds in the chalcogenide waveguide core, compared to the cladding, enable simultaneous guiding of light and sound in the same rib structure.

2.1 Fabrication of chalcogenide rib waveguides and racetrack cavity

Nonlinear optical phenomena such as four-wave mixing, Raman scattering, self- and cross-phase modulation have been studied in chalcogenide glass photonic chips for over two decades (Eggleton et al., 2011; Spälter et al., 2002). The fabrication techniques for developing chalcogenide photonic chips have been ever evolving to incorporate new functionalities and to achieve hybrid integration. However, some of the basic fabrication steps remain the same.

Chalcogenide (primarily As_2S_3) waveguide fabrication starts with the deposition of thin films of chalcogenide material on a SiO_2 substrate using either pulsed laser deposition (Huang et al., 2004; Youden et al., 1993), chemical vapor deposition (Huang et al., 2004), thermal evaporation (Choi et al., 2010; Hu et al., 2007a), or sputtering (Tan et al., 2010). After thin-film deposition, an annealing process is performed to relax strained chemical bonds and eliminate homopolar bonding thereby reducing optical loss and making the film more "bulk glass like" (Choi et al., 2013). A thin protective layer is also usually deposited to protect the As_2S_3 film from the alkaline developer of photoresist patterning (Choi et al., 2008a,b). The protective layer prevents alkaline developer attack and induced roughening of the film surface and subsequent increase in the propagation loss.

The first patterned chalcogenide-based waveguides were built in the 1970s. Ohmachi (1973) and AU-Ziegel et al. (1975) demonstrated acousto-optic light diffraction using a 1-D slab waveguide made of As_2S_3 deposited on lithium niobate. A laser beam was coupled in and out of the waveguide with a pair of GaP prism couplers. These waveguides could support only straight light propagation. In the late 1970s, the photo-darkening method was applied to the fabrication of curved waveguides (Zembutsu and Fukunishi, 1979). Optically induced changes in the chemical bonds of chalcogenide glasses modify the bandgap and refractive index. A photo-induced refractive index change was achieved by irradiating a thin chalcogenide glass film on the substrate with laser light near the bandgap. This direct writing method could induce an index change of about 0.03, which is large enough to use it as the waveguide core. In addition, the use of ion implantation for waveguide writing has been developed for higher refractive index contrasts (0.3–0.4) between core and cladding (DeCorby et al., 2005; Fick et al., 2002). Waveguides fabricated using these techniques have large mode size due to low index contrast, which results in smaller pump intensity and makes them inefficient for observing nonlinear optical effects.

Structured chalcogenide waveguides with high index contrast require photolithography using various fabrication methods. The first photolithographic method is the lift-off method (Hu et al., 2007b). Photolithography-based lift-off processes produce strip and rib type single-mode waveguides using the following process flow steps: (i) thick photoresist coating on a cladding substrate, (ii) light exposure (patterning), (iii) development, (iv) chalcogenide glass deposition, (v) lift-off (for strip-type waveguides), and additional chalcogenide glass deposition (for rib-type waveguides). The achieved propagation loss using the lift-off method was $<0.5\,dB/cm$ in $Ge_{23}Sb_7S_{70}$ waveguides (Hu et al., 2007b) and $3.2\,dB/cm$ in As_2S_3 waveguides (Hu et al., 2010). The surface morphology (after thermal reflow) of an As_2S_3 waveguide fabricated by the lift-off method is shown in Fig. 2A. The surface morphology was imaged with an atomic force microscope.

Due to the relatively low glass transition temperatures of chalcogenide glasses, chalcogenide waveguides can also be fabricated by thermal nanoimprint lithography (Han et al., 2010, 2011; Lian et al., 2009; Solmaz et al., 2008). Thermal nanoimprint technologies have the advantages of low cost, manufacturability, fast process, and no chemical use. Hard stamps (typically silicon) or soft stamps (usually PDMS) can both be used due to the relatively low glass transition temperatures of chalcogenide glasses. Soft stamps have the benefit of lower cost and simple replacement, and of being more compliant, providing potential for stamping on non-planar surfaces. Additionally, thermal nanoimprint also has the advantage of being able to

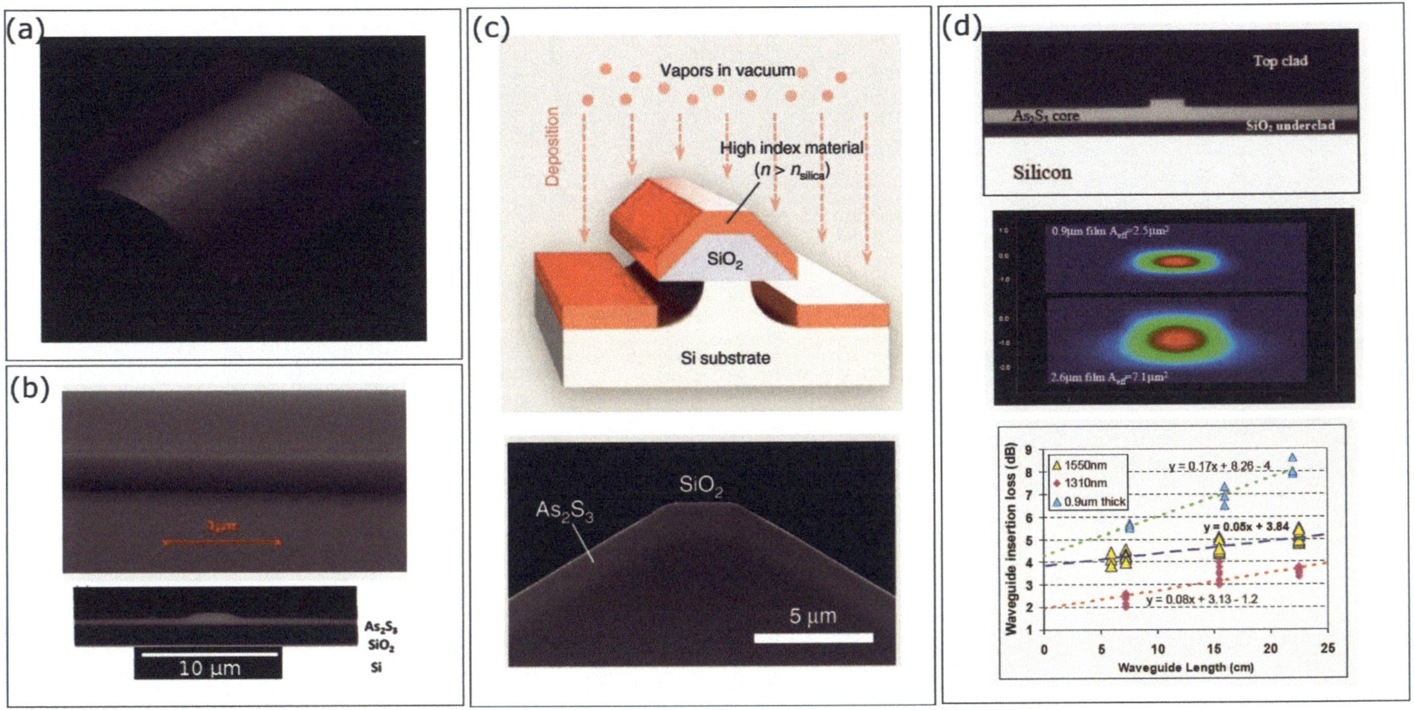

Fig. 2 Images of fabricated As_2S_3 waveguides. (A) Surface morphology of As_2S_3 chalcogenide waveguides fabricated by the lift-off method with thermal reflow (Hu et al., 2010). (B) top-down view and cross-sectional images of an As_2S_3 waveguide fabricated by thermal nanoimprint (Han et al., 2010). (C) Chalcogenide glass deposition on a pre-patterned cladding structure and the cross-sectional image of an As_2S_3 waveguide fabricated by the fabrication method without direct etching of waveguide core (Kim et al., 2020). (D) Cross-sectional image of an As_2S_3 waveguide fabricated by the direct etching method with CHF_3 gas, computed quasi-transverse electric mode field, and measured propagation loss for $2.5 \times 4\,\mu m$ and $0.9 \times 4\,\mu m$ waveguides at 1550nm and 1310nm (Hu et al., 2007b; Madden et al., 2007).

directly pattern rare earth doped chalcogenide films where reactive ion etching struggles due to non-volatile reaction compounds, this being of value for future higher level integration of pump lasers or amplifiers for example. The standard process flow for soft stamp techniques is: (i) fabrication of a mold master by photolithography and reactive ion etching, (ii) fabrication of a cast PDMS stamp from the master, (iii) deposition of a chalcogenide glass film on a substrate, (iv) heating the sample and stamp above the sample glass transition temperature and applying a uniform pressure to the stamp for a preset time, (v) rapidly cooling the sample, releasing the force and stamp (Han et al., 2010). The measured propagation losses by thermal nanoimprinting were 0.27 dB/cm in $As_{24}Se_{38}S_{38}$ waveguides and 0.52 dB/cm in As_2S_3 aveguides (Han et al., 2011). The top-down view and cross-sectional images of an As_2S_3 waveguide fabricated by the thermal nanoimprint method are shown in Fig. 2B. The top-down and cross-sectional images were obtained by scanning electron microscope microscopy and optical microscopy, respectively.

In 2020, a novel fabrication method without direct etching of waveguide core was demonstrated (Kang et al., 2017; Kim et al., 2020). Unlike the methods mentioned above, a cladding structure with exquisitely low surface roughness is fabricated first through photolithography and wet chemical etching, and then a chalcogenide glass film is deposited on the cladding. Through this method, the core structures have no discontinuity, reducing scattering from rough sidewalls. The measured propagation loss was 0.0162 dB/cm in the As_2S_3 glass waveguide, the lowest value reported to date (Morrison et al., 2017). The schematic of the chalcogenide glass deposition process on a pre-patterned cladding structure and the cross-sectional image of a chalcogenide waveguide fabricated by the method without direct etching of waveguide core are shown in Fig. 2C. The cross-sectional image was taken by scanning electron microscopy.

The methods mentioned above have additional challenges in manufacturing a complex optical system such as an optical splitter or an interferometer compared to directly patterning and etching the waveguide core. Accordingly, direct patterning and etching methods that can be developed into photonic integrated circuits have been thoroughly investigated (Choi et al., 2008a,b, 2010; Fan et al., 2020; Li et al., 2005; Madden et al., 2007). Dry etching and wet etching are possible, but due to the isotropic direction and low precision, fabricating complex photonic devices with wet etching is challenging (Riley et al., 2006). On the other hand, dry etching can directly transfer photolithographic patterns to a chalcogenide glass film with high anisotropy, high resolution and good controllability. Therefore, complex optical devices and even photonic integrated circuits can be manufactured this way. The As_2S_3 waveguide fabrication process using dry etching is as follows (Choi et al., 2010): deposition of As_2S_3 thin film on a cladding, annealing, covering the film with a thin protective layer (Choi et al., 2008a,b), photolithographic patterning, development, etching with plasma etcher with CHF3 gas (Choi et al., 2008b) or CF4 and O2 gas mix (Li et al., 2005), and coating a top cladding. As shown in Fig. 2D, the propagation loss measured in a 22.5 cm long As_2S_3 waveguide with a dimension of 2.5×4 um was 0.05 dB/cm at a wavelength of 1550 nm (Madden et al., 2007).

The process methodologies in References (Kim et al., 2020) and (Morrison et al., 2017) significantly reduced the propagation loss compared to other methods, which is clearly advantageous particularly in resonant type devices. Using the etch free method (Kim et al., 2020), the smallest optical losses to date in chalcogenide waveguides were achieved. But it is difficult to fabricate optical devices requiring precise spacing control such as directional couplers making this advantage harder to exploit with planar lithographically defined devices. As shown in Fig. 3A, coupling into a ring resonator can instead be attained using the flip-chip coupling method (Kim et al., 2020). Alternatively, using the direct etching method (Morrison et al., 2017), waveguides with sufficiently small optical loss and photonic devices requiring precise spacing control such as directional couplers were successfully fabricated. Fig. 3B illustrates a Brillouin laser comprising a chalcogenide ring resonator directional coupler, and silicon on insulator to chalcogenide waveguide transitions on a single hybrid integrated chip. Fig. 3C is the micrograph image of the Brillouin laser manufactured by the dry etching method. Although the Brillouin laser using the dry etching method requires 100 times higher threshold pump power than that achieved in Ref. (Kim et al., 2020), a monolithic photonic structure can be integrated on a single chalcogenide chip. Further reductions in the threshold can be achieved with improved lithography to reduce sidewall roughness. The development of these nanofabrication techniques has led to the demonstration of on-chip SBS and its application to areas ranging from signal processing to Brillouin lasing.

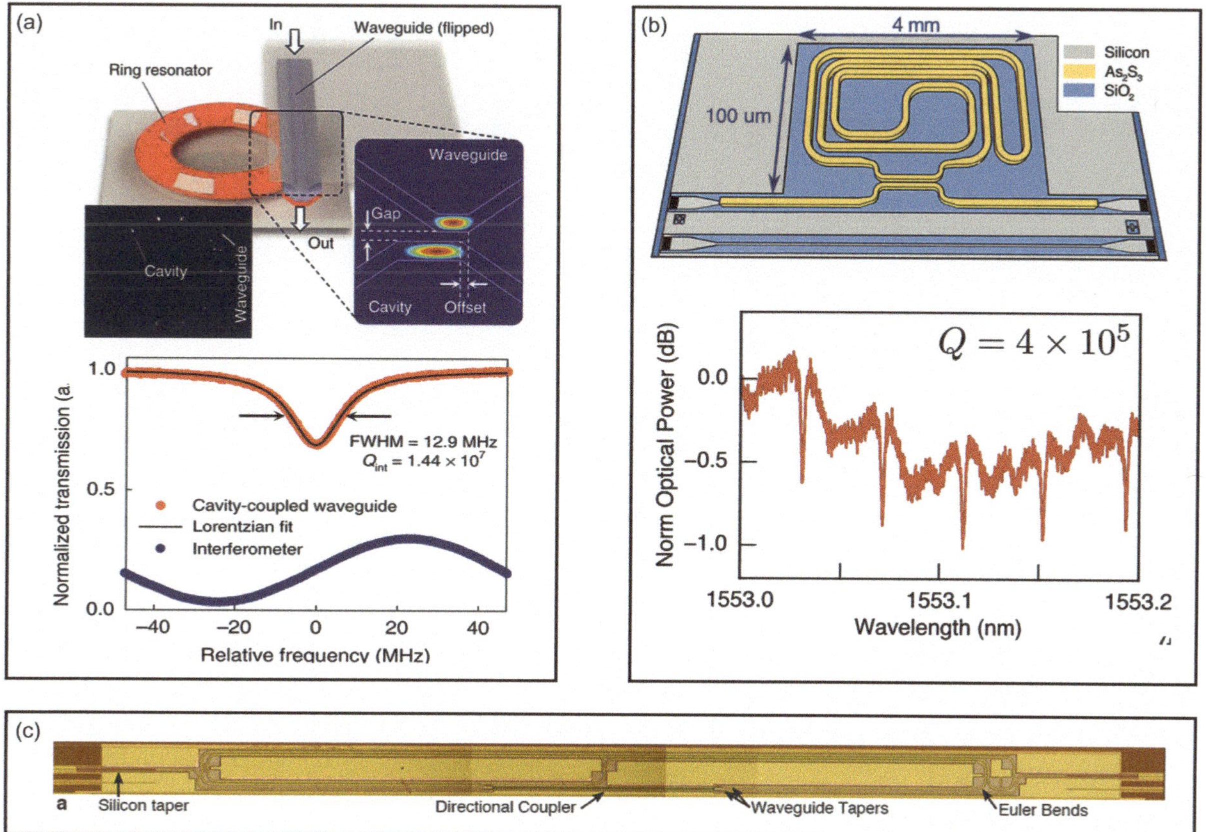

Fig. 3 See figure legend on next page.

3. On-chip SBS characterization

As discussed in chapter "Theoretical formalisms for stimulated Brillouin scattering" by Wolff et al., there are two configurations for characterizing Brillouin scattering in a medium. The simplest configuration is one where only a strong pump beam is sent through a medium and thermally generated acoustic phonons scatter the pump to generate the signals at the Stokes and anti-Stokes frequencies while satisfying energy and momentum conservation. This configuration is known as the generator configuration. The scattered signal in the generator configuration provides a direct measure of the frequency of the acoustic mode involved in the interaction with the pump. The drawback of this configuration is that high (Watt-level) continuous-wave (CW) pump power is needed to observe any appreciable scattered signal in most current chip-scale devices. Such high CW pump powers can cause a range of deleterious effects such as heating, optical damage, photorefractively driven mode conversion effects, which can significantly degrade performance. The use of a quasi-CW pump with the generator configuration circumvents these problems by reducing the average power while achieving large peak power. In the first on-chip SBS demonstration, a quasi-CW pump, consisting of 400 ns long optical pulses at a repetition rate of 25 kHz, was used to reduce the required average power to tens-of-mW while achieving a high peak power ~ few watts. The use of long pump pulses in these experiments ensures $t_{\text{pulse}} >> \tau_{phonon}$. The quasi-CW pump was obtained through intensity modulation of a CW laser followed by optical amplification using an Erbium-doped fiber amplifier. The appearance of a red-shifted signal in the backscattered optical spectra, shown in Fig. 4A, indicates the generation of a Stokes signal with an increase in the pump power (Pant et al., 2011b). As the pump power increases, the spontaneously generated Stokes signal experiences exponential gain causing

Fig. 3 Brillouin laser. (A) Schematic of the Brillouin laser and flip–chip coupling scheme. Both the ring resonator and waveguides are fabricated by the method without direct etching of waveguide core. The left inset shows the scattered light from the resonator taken by IR camera. The bottom figure shows the transmission spectrum with an intrinsic Q-factor of 14.4 million (Kim et al., 2020). (B) Schematic of the As_2S_3 ring resonator structure on SOI for Brillouin lasing, optical transmission of ring resonator fabricated by the direct etching method (Morrison et al., 2017). (C) Microscope images of the entire As_2S_3/SOI hybrid integrated circuit fabricated by the direct etching method (Morrison et al., 2017).

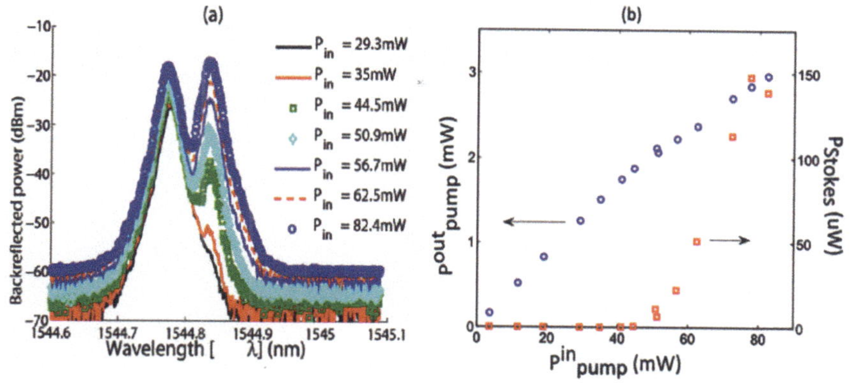

Fig. 4 Brillouin scattering spectrum. (A) Backscattered optical spectra from a chalcogenide photonic chip for different values of the input average pump power (B) Variation of the transmitted pump power and the backscattered Stokes power with the input pump power (Pant et al., 2011b).

a significant transfer of the pump power to the generated Stokes wave. At pump powers well above the Brillouin threshold, a significant part of the pump is backscattered resulting in saturation of the transmitted pump power, as seen from Fig. 4B. Brillouin scattering is, therefore, considered a nuisance for a number of applications, especially fiber-optic communications. No blue-shifted component is observed in Fig. 4A due to the loss experienced by the anti-Stokes wave. The shift of the Stokes signal wavelength from the pump wavelength gives a measure of the frequency of the acoustic mode involved in the Brillouin interaction. A Brillouin frequency of ∼7.7 GHz was observed for the chalcogenide photonic chip used in these experiments. The Brillouin generator configuration is typically used to study the elastic properties of a material by measuring the acoustic mode frequency and using it to calculate the acoustic wave speed V_a in the material. The generator configuration, when implemented in a resonator geometry, creates ultranarrow linewidth Brillouin lasers that are of utmost importance in coherent communications due to their low phase noise (Pant et al., 2011a; Li et al., 2013; Chauhan et al., 2021; Lee et al., 2012; Morrison et al., 2017). Many applications of SBS, however, use the pump-probe configuration, where a weak probe signal is propagated in the presence of a pump such that the pump remains undepleted. The probe signal experiences gain (loss) at the Stokes (anti-Stokes) frequency due to the SBS process.

Figs. 5A and B show the pump-probe characterization of the SBS gain profiles for the chalcogenide photonic chips fabricated using the vapor

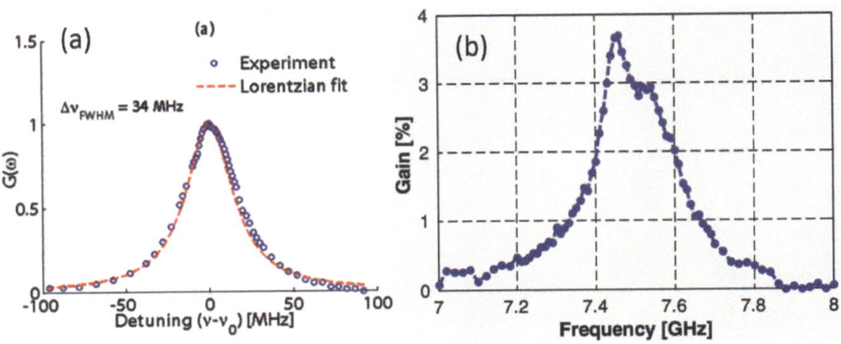

Fig. 5 Brillouin gain spectrum. (A) Brillouin gain profile for a chalcogenide rib waveguide fabricated using the vapor deposition method. The measured linewidth of 34 MHz is close to the linewidth obtained from the acoustic phonon lifetime for As_2S_3 (Pant et al., 2011b). (B) The measured Brillouin linewidth of 200 MHz for the laser written chalcogenide rib waveguide showing a broadening due to the excitation of different acoustic modes (Levy et al., 2012).

deposition (Pant et al., 2011b) and laser writing method (Levy et al., 2012), respectively. In pump-probe experiments, a probe is typically generated using electro-optic modulation of a CW laser of frequency ω_P with a radio frequency equal to the acoustic mode frequency Ω_B of the SBS medium. Electro-optic modulation creates the sidebands at frequencies $\omega_P \pm \Omega_B$. In this configuration, the laser frequency ω_P acts as the pump, and the modulation sidebands create the Stokes $(\omega_P - \Omega_B)$ and anti-Stokes waves $(\omega_P + \Omega_B)$. Tuning of the modulation frequency around the Brillouin shift varies the sideband amplitudes due to the Brillouin gain and loss resonances centered at the Stokes and anti-Stokes frequencies, respectively. Gain and loss profiles provide a measure of the inherent homogeneously broadened linewidth $\Delta\nu_B$ of the Brillouin resonance and the acoustic phonon lifetime τ_{phonon} through the relation $\Delta\nu_B = \frac{1}{\tau_{phonon}}$. However, if the Brillouin shift varies along the propagation length due to a variation in the cross-section, presence of more than one acoustic mode, and refractive index fluctuations, the linewidth will be broader than the actual linewidth determined by τ_{phonon}. The linewidths of the SBS gain profiles in Fig. 5A and B are 34 and 200 MHz, respectively. The 34 MHz linewidth is close to the inherent Brillouin linewidth in As_2S_3 glass. The broader linewidth of the laser written photonic chip could be due to a variation in the optical mode evolution. Conducting a pump-probe experiment by keeping the modulation frequency fixed at Ω_B and tuning the pump power gives a measure of the Brillouin gain coefficient g_B. The value of g_B is obtained by fitting an exponential function to the Stokes power vs pump power plot. In Ref. (Pant et al., 2011b), a value of 0.74×10^{-9} m/W was obtained for the Brillouin gain coefficient in the photonic chip. In the pump-probe configuration, the probe signal provides seed for stimulating the acoustic wave through interaction between the pump and the Stokes wave, which reduces the threshold pump power required for achieving a desired Brillouin gain. The SBS gain and loss resonance are also accompanied by sharp dispersion, which can be used to control the phase of the probe sideband. A number of applications such as slow- and fast-light, microwave photonic signal processing (Pant et al., 2014) require control over the amplitude and phase of a signal. Realizing these applications in a chip-scale platform is important for achieving photonic integration. SBS in these photonic integrated circuits has been harnessed to demonstrate a number of these applications, some of which are discussed in the next section.

4. On-chip SBS applications

4.1 Slow and fast light

Slow- and fast-light refers to a control of the group-velocity V_g of an optical signal propagating through a medium. The group-velocity is related to the group-index n_g, which depends on the dispersion slope $dn/d\omega$ in a medium. Large positive and negative values of $dn/d\omega$ occur at gain and loss resonances, respectively, which reduce or enhance the speed of light packets in the medium compared to vacuum. SBS slow- and fast-light has emerged as a potential candidate for tuning the delay of light pulses. The wavelength transparent nature and bandwidth reconfigurability, which allows delay control of different pulse width signals centered at different wavelengths, makes SBS an attractive technique for optical signal processing (Chin et al., 2010; Okawachi et al., 2005; Song and Hotate, 2007; Zhu et al., 2007). The resonant delay T_d created by SBS slow-light depends on the Brillouin pump power P_p, through the gain exponent G, according to $T_d = G/\Gamma_B$, where $G = g_B P_p L_{eff}/A_{eff}$ and $\Gamma_B = \frac{\Delta\nu_B}{2\pi}$. Varying the pump power, therefore, allows delay tuning. The group delay can be expressed in terms of the group index n_g as $T_d = Ln_g/c$, where $n_g = g_B P_p c/A_{eff}$ for the Brillouin slow-light. Large group delay is, therefore, possible with a long device even if the group-index is small, i.e., slow-light effect is weak, and the speed of light in the Brillouin medium does not reduce much compared to vacuum. Realizing large group delay in a chip, however, requires a device with large n_g to achieve a drastic reduction in the speed of light in the medium. Chalcogenide photonic chips, due to their large g_B and small A_{eff}, achieve large n_g and therefore have strong slow-light effects.

Fig. 6A shows the output pulses for a 100 ns long input pulse at different values of the Brillouin gain. Tuning of the gain through a variation in the pump power results in delay control. In (Pant et al., 2012), a large delay of 25 ns is achieved in a 6 cm long chalcogenide waveguide, which results in a group index of ~130. For the same length in a vacuum, the pulse will delay by 0.2 ns. On-chip Brillouin slow-light can, therefore, achieve 100 times smaller group-velocity compared to its value in a vacuum. The large n_g is obtained using a pump power of 300 mW. Achieving such large group index values using the same length of a silica fiber requires pump power of-the-order of a few watts. The delay vs. gain plot in Fig. 6B plots the measured delay for input pulse widths of 25 ns (squares) and 100 ns (circles) along with the theoretical resonance delay. The measured delay for 100 ns pulses agrees well with the theoretical delay whereas, the measured delay for a 25 ns pulse deviates from the theoretically predicted delay. For smaller pulse widths, the pulse bandwidth is larger than the Brillouin gain bandwidth, and therefore the gain experienced by the pulse is not the resonant gain as happens for a longer pulse. Further, for shorter pulses, the larger bandwidth causes pulse distortion due to the filtering of the pulse spectrum by the Brillouin gain profile, and the group-velocity dispersion. Broadening of the Brillouin gain profile reduces pulse distortion for pulse bandwidths larger

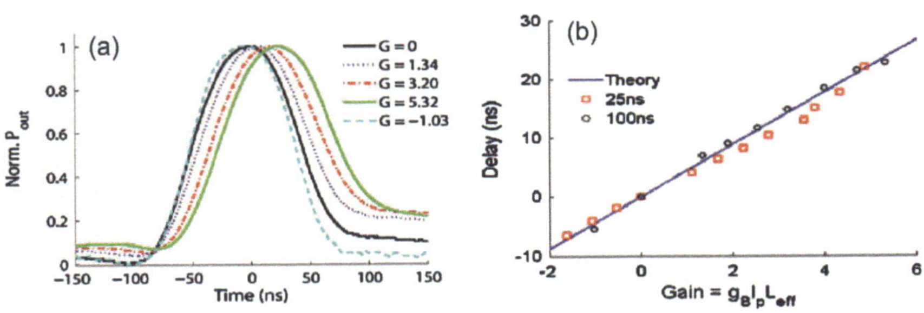

Fig. 6 (A) Normalized output pulses for a 100 ns long input pulse showing pulse delay for G > 0 and pulse advance for G < 0. (B) Comparison of the theoretical resonance delay at different gain values with the measured pulse delays for 25 and 100 ns long input pulses (Pant et al., 2012).

than the inherent Brillouin gain linewidth. Using a multi-line pump, gain profile was broadened to achieve true-time delay for a pulse bandwidth of ~100 MHz in a 24 cm long chalcogenide waveguide (Aryanfar et al., 2017). The SBS gain, phase and group delay profiles for a single Brillouin pump are shown in Figs. 7A–C, respectively. In (Aryanfar et al., 2017), a large on-chip gain of 52 dB was achieved using a single Brillouin pump, as shown in Fig. 7A, which is more than enough for a number of applications. Effect of pump profile broadening on the Brillouin gain, phase and group delay is shown in Fig. 7D–F. Increasing the pump bandwidth enables accommodation of large bandwidth pulses at the cost of reduced gain, as seen from Fig. 7D. The broader gain bandwidth reduces the phase slope and

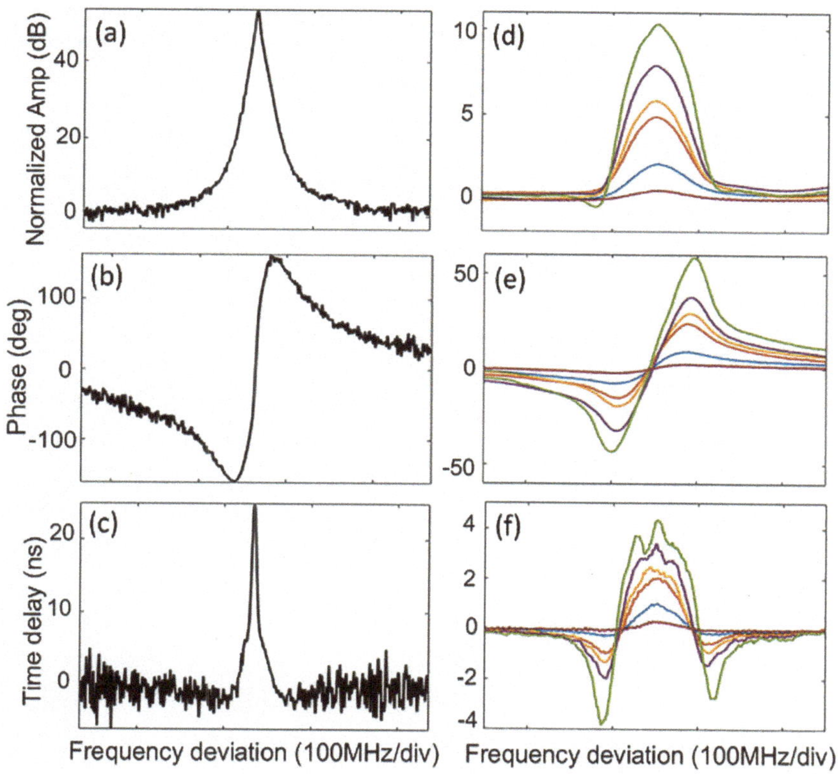

Fig. 7 Brillouin gain and time delay spectra. (A)–(C) Brillouin gain profile, phase response and group delay, respectively, for a 24 cm long chalcogenide waveguide using a single continuous-wave pump. A large on-chip gain of 52 dB was achieved at low pump powers. (D) Broadened gain profile and (E and F) corresponding phase response and group delay for a multi-line Brillouin pump. The measured Brillouin gain linewidth was extended to 100 MHz (Aryanfar et al., 2017).

group delay in Fig. 7E and F, respectively. Pump-probe configuration for applications such as slow- and fast-light use the same polarization for the pump and the probe signal. However, there are applications that require interaction between the acoustic grating written by optical waves in one polarization and probing of the grating by an optical signal in orthogonal polarization. Such a scheme is called a Brillouin dynamic grating (BDG) and is useful for applications in sensing, storage, optical signal processing, and spectral measurement.

4.2 Brillouin dynamic gratings

The Brillouin shift in a medium depends on the refractive index of the optical mode at the pump wavelength λ_P. For a birefringent medium, the refractive indices n_x and n_y for the orthogonally polarized pump beams are different, which results in different Brillouin shifts $\nu_{xB} = \frac{2n_x V_a}{\lambda_P}$ and $\nu_{yB} = \frac{2n_y V_a}{\lambda_P}$ for a given pump wavelength λ_P, as shown in the inset of Fig. 8A. However, a grating written by the pump and Stokes beam in one polarization is phase-matched to an orthogonally polarized probe of wavelength λ_{Probe}, where λ_{Probe} is obtained using the condition $\nu_{BDG} = \frac{2n_x V_a}{\lambda_P} = \frac{2n_y V_a}{\lambda_{Probe}}$. Therefore, a grating written using an x-polarized pump and Stokes wave can be read using a y-polarized probe λ_{Probe} to generate a read signal at $\nu_{read} = \nu_{Probe} - \nu_{BDG}$. Fig. 8A shows the process of grating writing in a photonic chip by the pump and Stokes, called pump1 and pump2, respectively, in x-polarization. In the absence of pump1, only the pump2 (Stokes) is present. When the pump1 is turned on, the pump2 signal undergoes amplification, representing the creation of an acoustic grating. The strong birefringence of the photonic chip in Ref. (Pant et al., 2013) causes a large separation ($\Delta\nu_{BS} \sim 130\,\text{MHz}$) between the Brillouin shifts of the orthogonally polarized pumps, as seen from the inset of Fig. 8A. The probe wavelength $\lambda_{Probe} = \frac{n_y}{n_x}\lambda_P$ is detuned ($\Delta\lambda$) from the pump1 wavelength according to $\Delta\lambda = \frac{\Delta\nu_{BS}}{\nu_{BDG}}\lambda_P$. The large birefringence of the photonic chip results in $\lambda_{Probe} \sim 1525\,\text{nm}$ and $|\Delta\lambda| \approx 25\,\text{nm}$ in these experiments. Fig. 8B shows the reading of the BDG by the probe. When only one of the pumps is on, no read signal appears at the frequency ν_{read}. In the presence of both the pumps, a BDG is created, and the probe scatters off from the BDG to generate a signal at ν_{read}, as shown in Fig. 8B. A large on-off ratio of 28 dB was observed for the read signal.

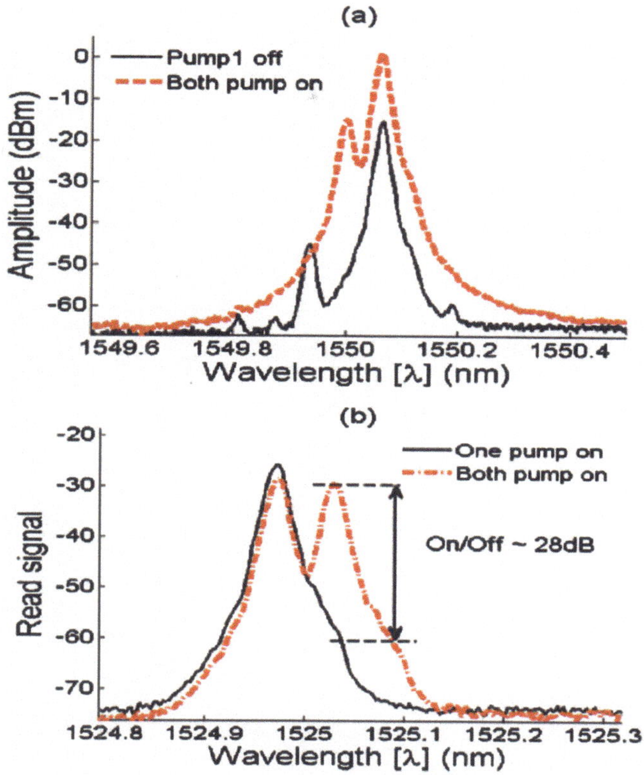

Fig. 8 Brillouin scattering spectrum. (A) Pump 2 (Stokes wave: black) spectrum in the absence of Pump 1. Amplification of Pump2 (red-dashed) in the presence of Pump1 showing the creation of a dynamic grating. Inset shows the radio-frequency spectrum demonstrating different Brillouin shifts for pump in two orthogonal polarizations. (B) Reading of the dynamic grating using a probe (solid black). Probe reflection from the dynamic grating when both pump1 and pump 2 are on (red-dashed). A read signal with an on-off ratio of 28 dB was observed (Pant et al., 2013).

The strength of the acoustic grating can be controlled by tuning the optical power of pump1 and pump2. Fig. 9A shows the spectral evolution of the read signal when the pump1 power is varied while keeping pump2 power in the write polarization and probe power in the read polarization fixed. The pump1 power was varied using an erbium-doped fiber amplifier that increased the noise floor in Fig. 9A. With an increase in the pump1 power, the grating strength increases due to the gain seen by pump2, which causes an increase in the read signal at frequency ν_{read}. The percentage ratio of the read signal power to the input probe power in Fig. 9B shows an exponential increase in the read signal power with pump1 power P_{p1}. The exponential

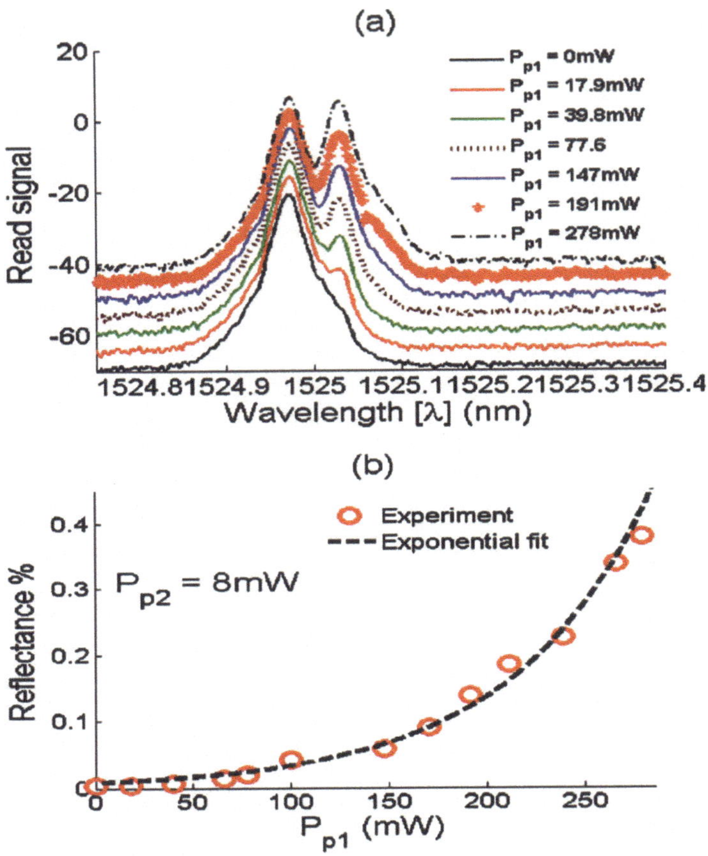

Fig. 9 Brillouin scattering spectrum. (A) Optical spectra showing an increase in the read signal at $\nu_{read}=\nu_{Probe}-\nu_{BDG}$ when the pump1 power is increased for a fixed probe power. (B) Read signal, plotted as a percentage of the probe power transferred to ν_{read}, showing an exponential increase with pump1 power (Pant et al., 2013).

increase in the read signal power shows the exponential increase in the gain seen by the pump2 due to the increased BDG strength. In these experiments, the probe wavelength λ_{Probe} was kept fixed at the wavelength that is phase-matched to the BDG, and the pump2 frequency was downshifted from pump1 exactly by the Brillouin shift for the write polarization. Detuning of the pump2 frequency from the resonance condition $\nu_{Pump2}=\nu_{Pump1}-\nu_{BDG}$, however, reduces the BDG strength, which results in a smaller read signal.

Fig. 10A shows the normalized read signal power as the pump2 frequency is detuned from the value $\nu_{Stokes}=\nu_{Pump1}-\nu_{BDG}$ (Pant et al., 2013).

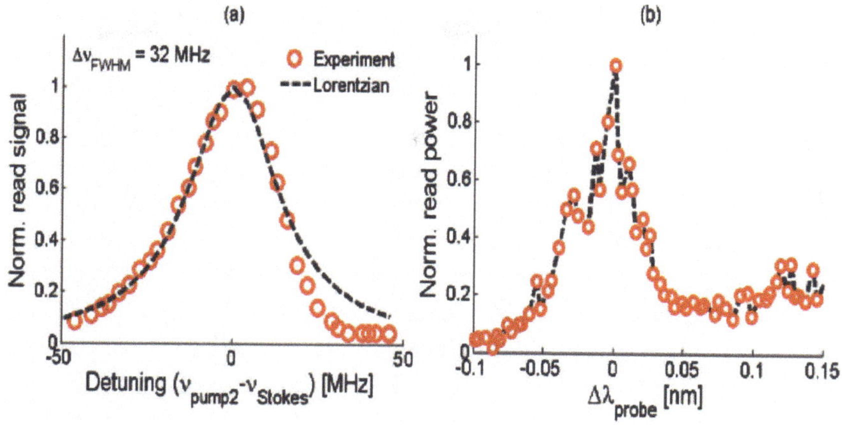

Fig. 10 Brillouin scattering spectrum. (A) Normalize read signal power at a fixed probe power when frequency of the pump2 is detuned from the Stokes frequency. The read signal follows the Brillouin gain profile as pump2 frequency is tuned around the Stokes frequency. (B) Normalized read signal power vs probe frequency detuning when the separation between pump1 and pump2 is kept at $\nu_{BDG} = \nu_{Pump1} - \nu_{Stokes}$ and powers of pump1, pump2 and probe are fixed (Pant et al., 2013).

The pump2 frequency is tuned by varying the radio frequency into the modulator away from ν_{BDG} while keeping the probe frequency fixed. This reduction in the read signal power shows that the grating strength is maximum when the pump2 frequency is at the Brillouin gain resonance. The read signal power drops to half of its maximum value at a detuning of $\Gamma_B/2$ (16 MHz) from the peak, which shows that the read signal power follows the Brillouin gain bandwidth when the pump2 frequency is tuned. Tuning the probe frequency while keeping the frequency separation between pump1 and pump2 at $\nu_{BDG} = \nu_{Pump1} - \nu_{Stokes}$ gives the linewidth of the BDG.

Fig. 10B shows the variation of the read signal power as the probe frequency is varied from its phase-matched value (Pant et al., 2013). The full-width at half-maximum linewidth $\Delta\nu_{BDG} = c/2nL$ of a BDG is inversely proportional to the device length L. For a 7 cm long chalcogenide (As_2S_3) rib waveguide, $\Delta\nu_{BDG}$ is ~1 GHz. The observed $\Delta\nu_{BDG}$ of 6 GHz for the photonic chip was nearly six times larger than the theoretical estimate. Variation in birefringence, cross-sectional area and other device parameters along the waveguide increase the $\Delta\nu_{BDG}$. Dynamic gratings are useful for a number of optical signal processing tasks such as storage, tunable delay, optical differentiation and integration, microwave photonic signal processing, and true-time reversal (Santagiustina et al., 2013). On-chip BDG allows photonic integration of these functionalities.

4.3 Ultrashort pulse generation using phase-locking

Ultrashort pulse sources with GHz repetition rates have applications in areas ranging from optical communications to spectroscopy. Short pulse widths ~ fs to ps can be obtained using mode-locked lasers, where phases of multiple cavity modes within the gain medium bandwidth are locked to create a periodic interference pattern in the time domain that results in a pulse train. Achieving short pulses with GHz repetition rates requires a small cavity length. On-chip Brillouin scattering can generate multiple Stokes components in the presence of feedback, which is provided by the Fabry-Perot cavity formed by index mismatch at the input and output facets with air. Feedback results in cascaded Brillouin scattering, where the first Stokes acts as the pump to generate a second Stokes once the first-Stokes power reaches the threshold and the process continues with higher-order Stokes. The four-wave mixing between the pump and multiple Stokes components gives rise to anti-Stokes frequencies, as shown in Fig. 11A (Büttner et al., 2014).

The photosensitivity of soft glass materials such as chalcogenides further enhances the efficiency of on-chip multi-Stokes generation due to the writing of a Hill-type grating (Hill et al., 1978). In (Büttner et al., 2014), presence of a Hill-type grating provided feedback to the first- and second-order Brillouin Stokes to reduce the Brillouin threshold for their excitation. Fig. 11B and C show the Brillouin pump and presence of Hill-type grating, respectively, which was written prior to the launch of the pump. By adjusting the Brillouin pump wavelength, the first- and second-order Stokes are enhanced due to the lowering of the threshold power for Stokes generation by the grating. Fig. 11D shows the backscattered spectra in the absence and presence of the Hill-type grating, demonstrating generation of Stokes components up to fourth-order in the presence of grating. Four-wave mixing of Stokes lines with the pump creates anti-Stokes lines up to third-order. In this demonstration (Büttner et al., 2014), a 300 ns long pulse created through electro-optic modulation was used as the Brillouin pump after optical amplification. The Brillouin comb lines, in general, are not equidistant due to the dependence of the Brillouin shift on the pump wavelength. However, pulling of frequency components due to the cavity resonance and Brillouin gain helps in maintaining a fixed frequency difference between these lines to form a comb. Four-wave mixing mediated phase-locking of Brillouin comb lines create ultrashort pulses, as shown in Fig. 12. The pulse trains measured at regular intervals are shown in Fig. 12A. The generated pulses at two different instants, I and II in Fig. 12A, are shown in Fig. 12B. Combining on-chip SBS with a Hill-type grating offers a simple-to-implement method for creating ultrashort pulses.

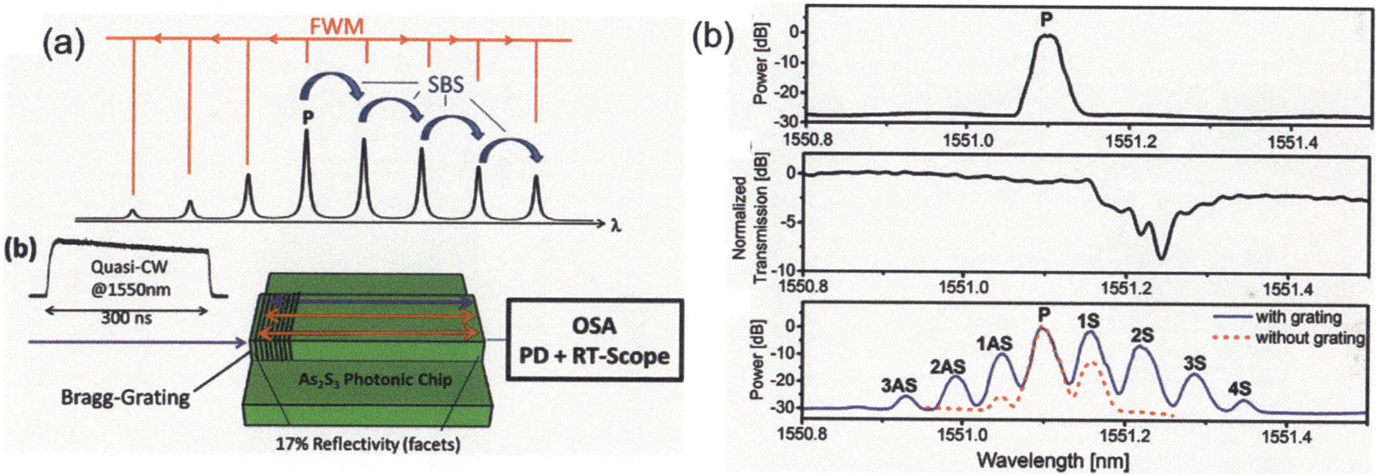

Fig. 11 Cascaded Brillouin scattering. (A) Schematic of cascaded Brillouin scattering in the presence of feedback and generation of four-wave mixing mediated anti-Stokes (top). Formation of a Hill-type grating in the chalcogenide photonic chip by a quasi-CW pump (bottom). (B, C) Pump and Hill-type grating spectrum in the photonic chip (D) Backscattered spectra in the absence and presence of the grating demonstrating lowering of Brillouin threshold and multi-Stokes generation in the presence of grating (Büttner et al., 2014).

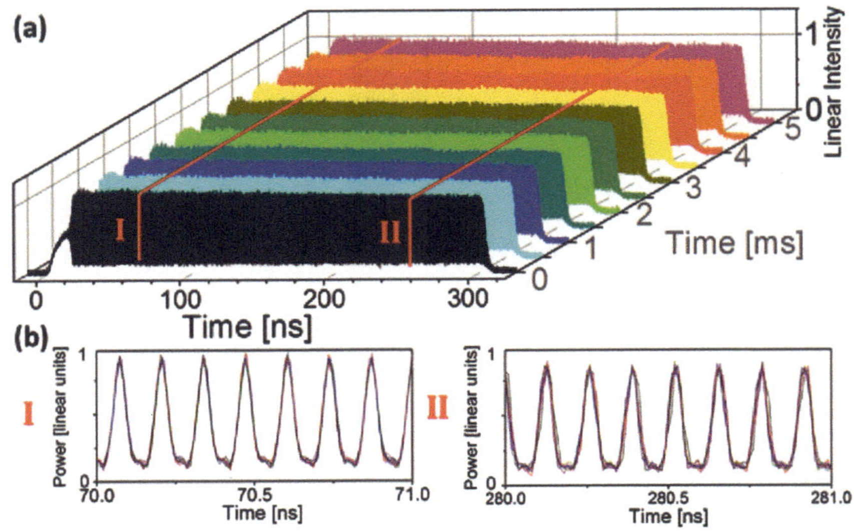

Fig. 12 (A) Ultrashort pulses obtained from phase-locking of multiple Stokes lines that are generated by a 300 ns long pump pulse. (B) Ultrashort pulses at two different time instants (I) and (II) within in the pump pulse duration (Büttner et al., 2014).

5. Integration with CMOS Platforms

Over the last decade, SBS has enabled applications in a variety of areas ranging from sensing, high coherence sources, microwave photonic signal processing, optical isolation, and slow−/fast-light-based delay tuning. It is expected to expand to an even more comprehensive range of applications by creating an SBS-based hybrid microwave-sound-photonics systems platform (Kittlaus et al., 2018). Combining strong on-chip SBS generation with a CMOS (Complementary Metal-Oxide Semiconductor) compatible platform, which is capable of generating, routing, modulating, processing, and sensing light, will herald a new era in photonic signal processing with unprecedented performance.

Silicon photonics directly repurposes the maturity, know-how and existing infrastructure of silicon CMOS processing technology to photonics and optoelectronics. Fabrication methods, including doping, etching, patterning, and bonding, have been well developed. Creating optoelectronic devices using SBS as a functional component of active photonic circuits could develop a new and powerful information processing platform combined with integrated microwave photonics. To this end, SBS optical processing in a CMOS compatible platform has been demonstrated using silicon-on-insulator (SOI) device architectures with various structures (Kittlaus et al., 2016; Shin et al., 2013, 2015; Van Laer et al., 2015). However, since the nonlinear loss of silicon is large and the Brillouin scattering nonlinear coefficient of bulk silicon material is smaller than those of silica materials, silicon is not an optimal material for strong Brillouin scattering (Eggleton et al., 2019). On the other hand, chalcogenide glass has a high refractive index, low nonlinear absorption loss, large Brillouin scattering nonlinearity, and can easily implement strong backward SBS on-chip. However, compared with silicon-based optical integrated circuits, the progress in development of chalcogenide glass-based integrated circuits is still slow, and there are difficulties to overcome for commercial mass production. Hybrid integration, where chalcogenide devices are developed on a CMOS-compatible platform, is, therefore, the way forward to harness the advantages of two material systems for achieving unprecedented performance and scalability.

A method of integrating two materials on a single chip was proposed in (Morrison et al., 2017; Zarifi et al., 2018). In this hybrid integration approach, conventional silicon-grating-coupler-based fiber interfaces are used for in- and out-of-chip light coupling, and a combined silicon/chalcogenide taper is used to provide a low-loss and low reflection transition between the silicon and chalcogenide glass sections of the circuit. This hybrid integrated approach is possible because chalcogenide glass can easily be deposited on conventional silicon wafers by various deposition methods (Choi et al., 2010; Hu et al., 2007a; Huang et al., 2004; Tan et al., 2010; Youden et al., 1993). When such a soft glass component is deposited, it can readily permeate into existing silicon structures, enabling structures such as slot waveguides (Mirnaziry et al., 2016), and large Brillouin gain is expected in such a hybrid system.

The primary sequence of the silicon and chalcogenide glass hybrid optical integrated circuit process begins with fabricating a silicon section, including a grating coupler and a taper (Morrison et al., 2017). The grating coupler provides the coupling of light from the optical fiber to the optical waveguide. The taper section is for adiabatically converting the fundamental mode of the silicon optical waveguide into the fundamental mode of the glass optical waveguide. In the implemented hybrid integrated circuit, the width of the silicon nanowire is adiabatically reduced to 150 nm before interfacing with the 690 nm thick chalcogenide (As_2S_3) region, and the conversion loss in the tapering section is about 0.1 dB, according to the simulation results. After processing the silicon section, lithographic processing was used to remove the protective polymer layer in a localized area where the chalcogenide device was to be fabricated, and then deposition of a 680 nm thick As_2S_3 layer was performed on the localized area using a shadow mask process. The As_2S_3 optical waveguide patterning was then performed using an e-beam lithography and RIE process, and 1 μm thick silica cladding was sputtered after etching.

The fabrication technique has demonstrated hybrid integration of a silicon-based platform and soft glass (Morrison et al., 2017; Zarifi et al., 2018). Several As_2S_3-silicon hybrid devices have been fabricated, including an on-chip hybrid circuit containing an As_2S_3 spiral waveguide and a chip with a small planar As_2S_3 ring resonator. As shown in Fig. 13A, the spiral waveguide has a total device length of 5.8 cm with a compact size of 0.1 mm × 4 mm. The numbers indicate on-chip optical components: 1. Silicon grating couplers with tapers to 450 nm × 220 nm nanowires; 2. Silicon nanowire taper region with As_2S_3 overlay waveguide; 3. As_2S_3 waveguide lead into the hybrid structure; 4. Spiral waveguide formed out of As_2S_3; 5. Alignment markers formed in the silicon layer for patterning the As_2S_3 structures; 6. Reference silicon structures existing on the same chip. SBS interaction occurs in the As_2S_3 soft glass waveguide while light coupling in and out of the waveguide was provided via silicon grating couplers and combined silicon/chalcogenide tapers. A silicon nanowire taper region with an As_2S_3 overlay waveguide is used to transmit light from/to a silicon grating coupler to/from the chalcogenide glass section with low loss and low reflection. Fig. 13B shows the SEM image of the end of the silicon taper before cladding deposition. The total insertion loss was on the order of 0.1 dB for transmission into the fundamental mode of the As_2S_3 waveguide (Morrison et al., 2017, Zarifi et al., 2018). Figs. 13C and D are the SEM image and schematic of chalcogenide region cross section with silica cladding, respectively, and Fig. 13E shows the optical mode simulation of fundamental TE mode of 1.9 μm wide As_2S_3 waveguide. The Brillouin amplification spectrum through the waveguide was measured with the pump-probe experiment, as shown in Fig. 13F. The results show the pump power dependence of SBS gain. As the pump power increases, the peak gain increases nearly linearly under the experimental conditions, as shown in Fig. 13G. In addition, reducing nonlinear losses is important in integrated silicon photonic circuits. Since the nonlinear absorption of soft glass is much

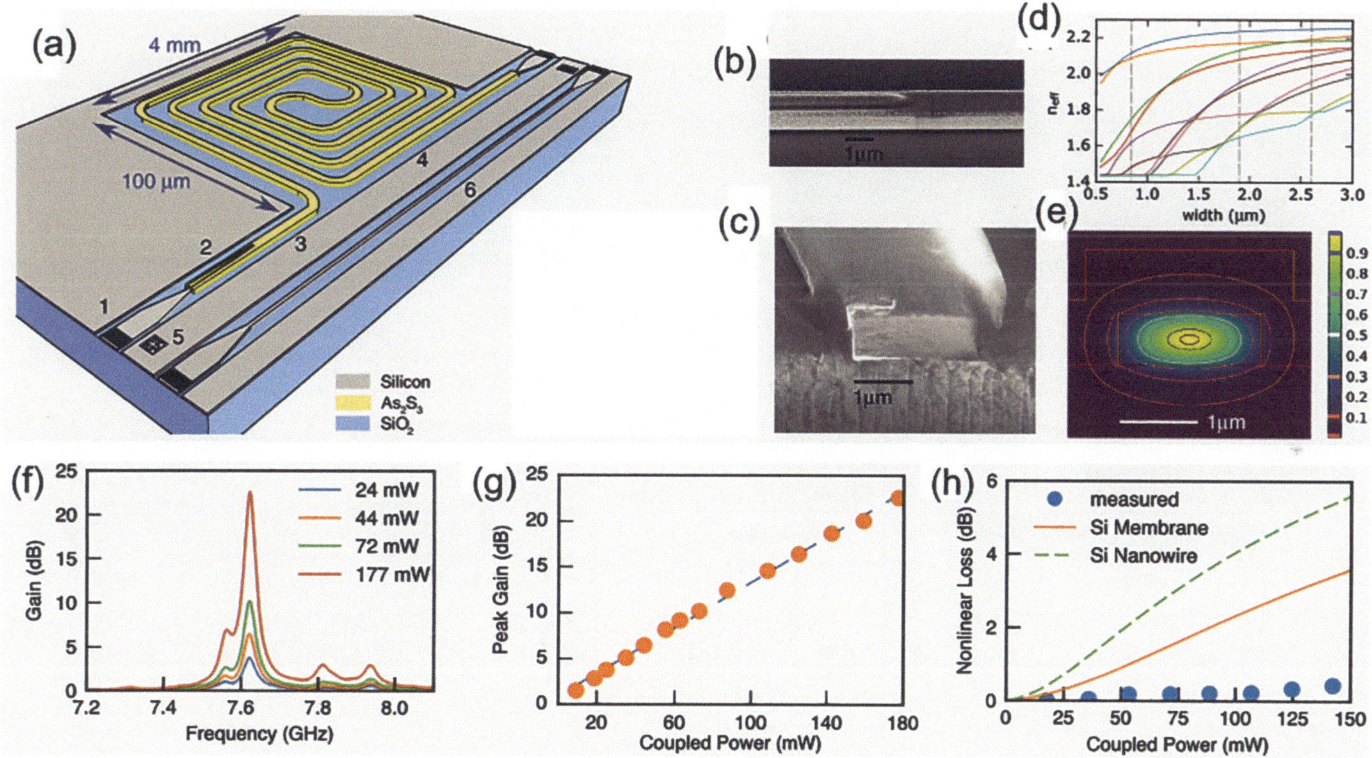

Fig. 13 Hybrid silicon/chalcogenide SBS waveguide. (A) Schematic of the hybrid circuit. (B) SEM image of the end of the silicon taper before cladding deposition. (C) SEM image of chalcogenide region cross section with silica cladding. (D) Schematic cross section of waveguide in chalcogenide-only region. (E) Optical mode simulation of fundamental TE mode of 1.9 µm wide As_2S_3 waveguide. (F) SBS spectrum for various pump powers. (G) Peak gain values up to 180 mW coupled pump power with fit. (H) Nonlinear loss comparison of the hybrid system, silicon nanowire, and silicon membrane (Morrison et al., 2017, Zarifi et al., 2018).

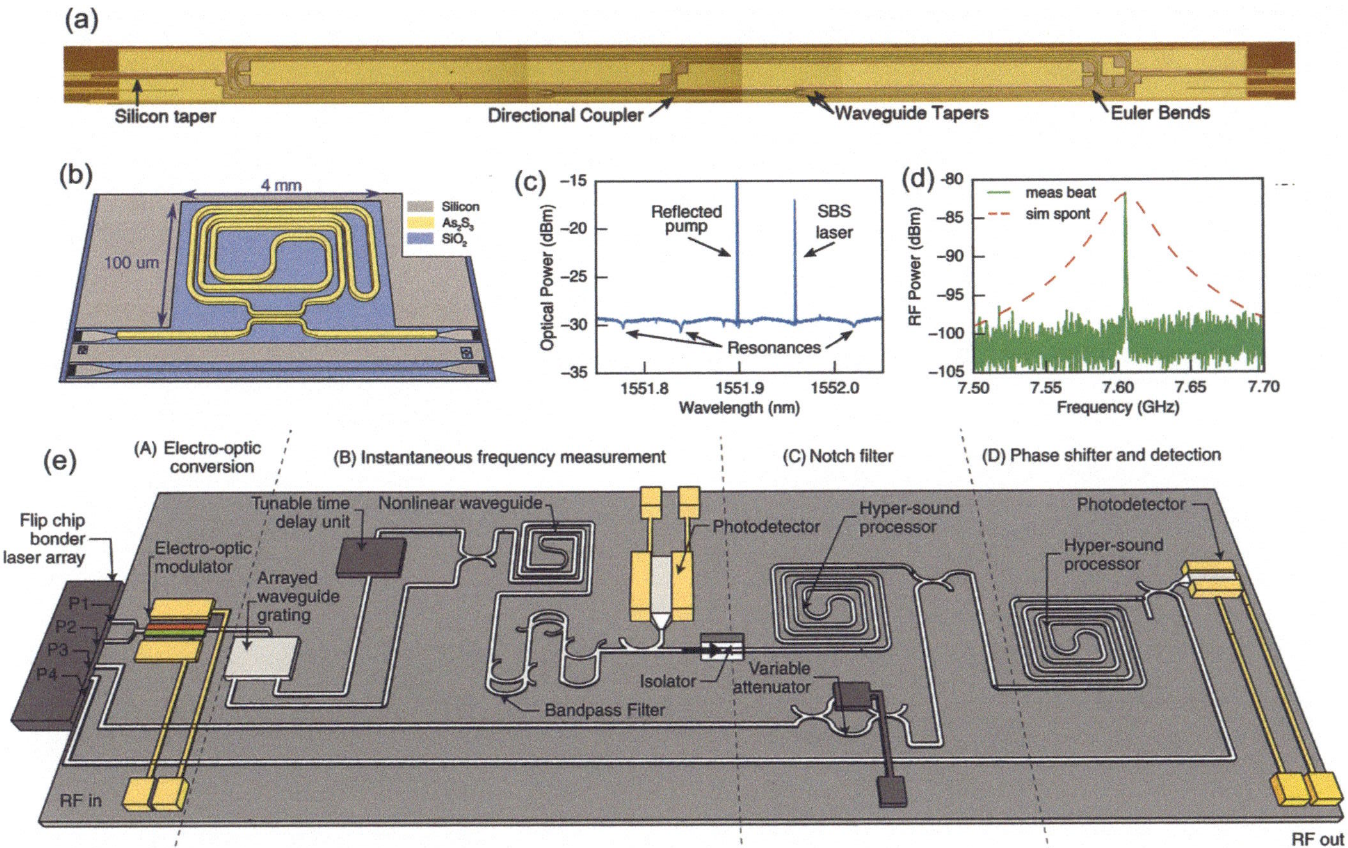

Fig. 14 See figure legend on opposite page.

lower than that of silicon optical waveguides, the measured nonlinear absorption of As_2S_3 waveguides is much lower than that of simulated silicon waveguides, as shown in Fig. 13H.

Also, Brillouin laser generation was observed in a planar ring resonator using an As_2S_3-silicon hybrid system (Morrison et al., 2017). Fig. 14A and B show the fabricated hybrid structure and its schematic, respectively. The RF beat signal spectrum between the generated SBS laser light and the reflected pump (Fig. 14C) shows that the linewidth of the generated laser light is 5 MHz, which is much narrower than the natural lifetime of the resonator of 40 MHz (Fig. 14D), confirming that the resonator emission exceeded the lasing threshold.

Fig. 14E shows a conceptual illustration of a chip-scale integrated microwave/RF processor for broadband communications, consisting of four sections. Each section provides unique functions, including electro-optic conversion, frequency measurement, filtering, and phase shifting (Eggleton et al., 2019). Although significant challenges remain for the realization of this new technology and its adoption in communications and sensing applications, the technology for mass production and application of integrated chips is advancing at tremendous speed. The use of a hybrid system with silicon and soft glass will enable the realization of fully integrated active SBS devices such as an integrated optoelectronic oscillator (Merklein et al., 2016), microwave photonic filter (Liu et al., 2016), on-chip optical gyroscope (Li et al., 2017), quantum memory (Kashkanova et al., 2017), Brillouin cooling (Chen et al., 2016), and Brillouin optical isolator based on parity-time symmetry breaking (Ma et al., 2020) and provide new opportunities for basic research using the interaction of light and sound.

Fig. 14 Brillouin lasing in planar As_2S_3 resonator and a chip-scale integrated microwave/RF processor for broadband communications. (A) Microscope images (not to scale) highlighting key device components. (B) Schematic of the hybrid ring resonator structure. (C) Brillouin lasing while monitoring the resonance position. Both the pump and generated Stokes are aligned to cavity resonances. (D) RF beat of the back-reflected pump and lasing signal. The measured linewidth was less than 5 MHz, significantly narrower than the natural lifetime of 40 MHz, confirming the lasing above the threshold power (Morrison et al., 2017). (E) A chip-scale integrated microwave/RF processor for broadband communications. Incoming RF signals are (A) converted to photons and (B) spectrally analyzed. The desired signal is (C) extracted and (D) processed before conversion back to RF. Functions (B), (C) and (D) all require strong on-chip optoacoustic interactions (Eggleton et al., 2019).

References

Abedin, K.S., 2006. Stimulated Brillouin scattering in single-mode tellurite glass fiber. Opt. Express 14, 11766–11772.

Aryanfar, I., Marpaung, D., Choudhary, A., Liu, Y., Vu, K., Choi, D.-Y., Ma, P., Madden, S., Eggleton, B.J., 2017. Chip-based Brillouin radio frequency photonic phase shifter and wideband time delay. Opt. Lett. 42, 1313–1316.

Au-Ziegel, V.V., Au-Litvinenko, A.A., Au-Ulianov, G.K., Au-Chalabian, G.A., 1975. Ultrasonic dispersive waveguide with a layer of chalcogenide glass on lithium niobate. Sov. Phys. Acous., 21.

Büttner, T.F.S., Merklein, M., Kabakova, I.V., Hudson, D.D., Choi, D.-Y., Luther-Davies, B., Madden, S.J., Eggleton, B.J., 2014. Phase-locked, chip-based, cascaded stimulated Brillouin scattering. Optica 1, 311–314.

Chauhan, N., Isichenko, A., Liu, K., Wang, J., Zhao, Q., Behunin, R.O., Rakich, P., Jayich, A.M., Fetig, C., Hoyt, C.W., Blumenthal, D.J., 2021. Visible light photonic integrated Brillouin laser. Nat. Commun. 12 (1), 4685. https://doi.org/10.1038/s41467-021-24926-8.

Chen, Y.-C., Kim, S., Bahl, G., 2016. Brillouin cooling in a linear waveguide. New J. Phys. 18, 115004.

Chin, S., Thévenaz, L., Sancho, J., Sales, S., Capmany, J., Berger, P., Bourderionnet, J., Dolfi, D., 2010. Broadband true time delay for microwave signal processing, using slow light based on stimulated Brillouin scattering in optical fibers. Opt. Express 18, 22599–22613.

Choi, D.-Y., Madden, S., Rode, A., Wang, R., Bulla, D., Luther-Davies, B., 2008a. A protective layer on As2S3 film for photo-resist patterning. J. Non Cryst. Solids 354, 5253–5254.

Choi, D.-Y., Madden, S., Rode, A., Wang, R., Luther-Davies, B., 2008b. Dry etching characteristics of amorphous As2S3 film in CHF3 plasma. J. Appl. Phys. 104, 113305.

Choi, D., Madden, S., Bulla, D.A., Wang, R., Rode, A., Luther-Davies, B., 2010. Submicrometer-thick low-loss As_2S_3 planar waveguides for nonlinear optical devices. IEEE Photonics Technol. Lett. 22, 495–497.

Choi, D.-Y., Wade, A., Madden, S., Wang, R., Bulla, D., Luther-Davies, B., 2013. Photo-induced and thermal annealing of chalcogenide films for waveguide fabrication. Physics Procedia 48, 196–205.

Decorby, R.G., Ponnampalam, N., Pai, M.M., Nguyen, H.T., Dwivedi, P.K., Clement, T.J., Haugen, C.J., Mcmullin, J.N., Kasap, S.O., 2005. High index contrast waveguides in chalcogenide glass and polymer. IEEE J. Sel. Top. Quantum Electron. 11, 539–546.

Eggleton, B.J., Luther-Davies, B., Richardson, K., 2011. Chalcogenide photonics. Nat. Photon. 5, 141–148.

Eggleton, B.J., Poulton, C.G., Pant, R., 2013. Inducing and harnessing stimulated Brillouin scattering in photonic integrated circuits. Adv. Opt. Photonics 5, 536–587.

Eggleton, B.J., Poulton, C.G., Rakich, P.T., Steel, M.J., Bahl, G., 2019. Brillouin integrated photonics. Nat. Photon. 13, 664–677.

Fan, Z., Yan, K., Zhang, L., Qin, J., Chen, J., Wang, R., Shen, X., 2020. Design and fabrication of As2Se3 chalcogenide waveguides with low optical losses. Appl. Optics 59, 1564–1568.

Fick, J., Nicolas, B., Rivero, C., Elshot, K., Irwin, R., Richardson, K.A., Fischer, M., Vallée, R., 2002. Thermally activated silver diffusion in chalcogenide thin films. Thin Solid Films 418, 215–221.

Gyger, F., Liu, J., Yang, F., He, J., Raja, A.S., Wang, R.N., Bhave, S.A., Kippenberg, T.J., Thévenaz, L., 2020. Observation of stimulated Brillouin scattering in silicon nitride integrated waveguides. Phys. Rev. Lett. 124, 013902.

Han, T., Madden, S., Bulla, D., Luther-Davies, B., 2010. Low loss chalcogenide glass waveguides by thermal nano-imprint lithography. Opt. Express 18, 19286–19291.

Han, T., Madden, S., Debbarma, S., Luther-Davies, B., 2011. Improved method for hot embossing As2S3 waveguides employing a thermally stable chalcogenide coating. Opt. Express 19, 25447–25453.

Hu, J., Tarasov, V., Agarwal, A., Kimerling, L., Carlie, N., Petit, L., Richardson, K., 2007a. Fabrication and testing of planar chalcogenide waveguide integrated microfluidic sensor. Opt. Express 15, 2307–2314.

Hu, J., Tarasov, V., Carlie, N., Feng, N.-N., Petit, L., Agarwal, A., Richardson, K., Kimerling, L., 2007b. Si-CMOS-compatible lift-off fabrication of low-loss planar chalcogenide waveguides. Opt. Express 15, 11798–11807.

Hill, K.O., Fujii, Y., Johnson, D.C., Kawasaki, B.S., 1978. Photosensitivity in optical fiber waveguides: application to reflection filter fabrication. Appl. Phys. Lett. 32 (10), 647–649. https://doi.org/10.1063/1.89881.

Hu, J., Feng, N.-N., Carlie, N., Petit, L., Agarwal, A., Richardson, K., Kimerling, L., 2010. Optical loss reduction in high-index-contrast chalcogenide glass waveguides via thermal reflow. Opt. Express 18, 1469–1478.

Huang, C.C., Hewak, D.W., Badding, J.V., 2004. Deposition and characterization of germanium sulphide glass planar waveguides. Opt. Express 12, 2501–2506.

Hwang, J., Kim, D.-G., Han, S., Jeong, D., Lee, Y.-H., Choi, D.-Y., Lee, H., 2021. Supercontinuum generation in As2S3 waveguides fabricated without direct etching. Opt. Lett. 46, 2413–2416.

Ippen, E.P., Stolen, R.H., 1972. Stimulated Brillouin scattering in optical fibers. Appl. Phys. Lett. 21, 539–541.

Kang, G., Krogstad, M.R., Grayson, M., Kim, D.-G., Lee, H., Gopinath, J.T., Park, W., 2017. High quality chalcogenide-silica hybrid wedge resonator. Opt. Express 25, 15581–15589.

Kashkanova, A.D., Shkarin, A.B., Brown, C.D., Flowers-Jacobs, N.E., Childress, L., Hoch, S.W., Hohmann, L., Ott, K., Reichel, J., Harris, J.G.E., 2017. Superfluid Brillouin optomechanics. Nat. Phys. 13, 74–79.

Kim, H., Shin, H., 2019. Large on-chip Brillouin net amplification in silicon-based nanophotonics. AIP Advances 9, 125032.

Kim, D.-G., Han, S., Hwang, J., Do, I.H., Jeong, D., Lim, J.-H., Lee, Y.-H., Choi, M., Lee, Y.-H., Choi, D.-Y., Lee, H., 2020. Universal light-guiding geometry for on-chip resonators having extremely high Q-factor. Nat. Commun. 11, 5933.

Kittlaus, E.A., Shin, H., Rakich, P.T., 2016. Large Brillouin amplification in silicon. Nat. Photon. 10, 463–467.

Kittlaus, E.A., Kharel, P., Otterstrom, N.T., Wang, Z., Rakich, P.T., 2018. RF-photonic filters via on-chip photonic–phononic emit–receive operations. J. Light. Technol. 36, 2803–2809.

Kobyakov, A., Sauer, M., Chowdhury, D., 2010. Stimulated Brillouin scattering in optical fibers. Adv. Opt. Photonics 2, 1–59.

Lee, H., Chen, T., Li, J., Yang, K.Y., Jeon, S., Painter, O., Vahala, K.J., 2012. Chemically etched ultrahigh-Q wedge-resonator on a silicon chip. Nat. Photon. 6 (6), 369–373. https://doi.org/10.1038/nphoton.2012.109.

Lee, K.K., Lim, D.R., Luan, H.-C., Agarwal, A., Foresi, J., Kimerling, L.C., 2000. Effect of size and roughness on light transmission in a Si/SiO2 waveguide: experiments and model. Appl. Phys. Lett. 77, 1617–1619.

Lee, H., Chen, T., Li, J., Painter, O., Vahala, K.J., 2012. Ultra-low-loss optical delay line on a silicon chip. Nat. Commun. 3, 867.

Levy, S., Lyubin, V., Klebanov, M., Scheuer, J., Zadok, A., 2012. Stimulated Brillouin scattering amplification in centimeter-long directly written chalcogenide waveguides. Opt. Lett. 37, 5112–5114.

Li, J., Lee, H., Vahala, K.J., 2013. Microwave synthesizer using an on-chip Brillouin oscillator. Nat. Commun. 4, 2097.

Li, W., Ruan, Y., Luther-Davies, B., Rode, A., Boswell, R., 2005. Dry-etch of As2S3 thin films for optical waveguide fabrication. J. Vac. Sci. Technol. A 23, 1626–1632.

Li, J., Suh, M.-G., Vahala, K., 2017. Microresonator Brillouin gyroscope. Optica 4, 346–348.

Lian, Z.G., Pan, W., Furniss, D., Benson, T.M., Seddon, A.B., Kohoutek, T., Orava, J., Wagner, T., 2009. Embossing of chalcogenide glasses: monomode rib optical waveguides in evaporated thin films. Opt. Lett. 34, 1234–1236.

Liu, Y., Marpaung, D., Choudhary, A., Eggleton, B.J., 2016. Lossless and high-resolution RF photonic notch filter. Opt. Lett. 41, 5306–5309.

Ma, J., Wen, J., Ding, S., Li, S., Hu, Y., Jiang, X., Jiang, L., Xiao, M., 2020. Chip-based optical isolator and nonreciprocal parity-time symmetry induced by stimulated Brillouin scattering. Laser Photonics Rev. 14, 1900278.

Madden, S.J., Choi, D.Y., Bulla, D.A., Rode, A.V., Luther-Davies, B., Ta'eed, V.G., Pelusi, M.D., Eggleton, B.J., 2007. Long, low loss etched As2S3 chalcogenide waveguides for all-optical signal regeneration. Opt. Express 15, 14414–14421.

Merklein, M., Stiller, B., Kabakova, I.V., Mutugala, U.S., Vu, K., Madden, S.J., Eggleton, B.J., Slavík, R., 2016. Widely tunable, low phase noise microwave source based on a photonic chip. Opt. Lett. 41, 4633–4636.

Mirnaziry, S.R., Wolff, C., Steel, M.J., Eggleton, B.J., Poulton, C.G., 2016. Stimulated Brillouin scattering in silicon/chalcogenide slot waveguides. Opt. Express 24, 4786–4800.

Morrison, B., Casas-Bedoya, A., Ren, G., Vu, K., Liu, Y., Zarifi, A., Nguyen, T.G., Choi, D.-Y., Marpaung, D., Madden, S.J., Mitchell, A., Eggleton, B.J., 2017. Compact Brillouin devices through hybrid integration on silicon. Optica 4, 847–854.

Ohmachi, Y., 1973. Acousto-optical light diffraction in thin films. J. Appl. Phys. 44, 3928–3933.

Okawachi, Y., Bigelow, M.S., Sharping, J.E., Zhu, Z., Schweinsberg, A., Gauthier, D.J., Boyd, R.W., Gaeta, A.L., 2005. Tunable all-optical delays via Brillouin slow light in an optical Fiber. Phys. Rev. Lett. 94, 153902.

Pant, R., Li, E., Choi, D.-Y., Poulton, C., Stephen, M., Luther-Davies, B., Eggleton, B., 2011a. Cavity enhanced stimulated Brillouin scattering in an optical chip for multiorder Stokes generation. Opt. Lett. 36, 3687–3689. https://doi.org/10.1364/OL.36.003687.

Pant, R., Poulton, C.G., Choi, D.-Y., Mcfarlane, H., Hile, S., Li, E., Thevenaz, L., Luther-Davies, B., Madden, S.J., Eggleton, B.J., 2011b. On-chip stimulated Brillouin scattering. Opt. Express 19, 8285–8290.

Pant, R., Byrnes, A., Poulton, C.G., Li, E., Choi, D.-Y., Madden, S., Luther-Davies, B., Eggleton, B.J., 2012. Photonic-chip-based tunable slow and fast light via stimulated Brillouin scattering. Opt. Lett. 37, 969–971.

Pant, R., Li, E., Poulton, C.G., Choi, D.-Y., Madden, S., Luther-Davies, B., Eggleton, B.J., 2013. Observation of Brillouin dynamic grating in a photonic chip. Opt. Lett. 38, 305–307.

Pant, R., Marpaung, D., Kabakova, I.V., Morrison, B., Poulton, C.G., Eggleton, B.J., 2014. On-chip stimulated Brillouin scattering for microwave signal processing and generation. Laser Photonics Rev. 8, 653–666.

Poulton, C.G., Pant, R., Eggleton, B.J., 2013. Acoustic confinement and stimulated Brillouin scattering in integrated optical waveguides. J. Opt. Soc. Am. B 30, 2657–2664.

Riley, B.J., Sundaram, S.K., Johnson, B.R., Saraf, L.V., 2006. Summary of Chalcogenide Glass Processing: Wet-Etching and Photolithography. Pacific Northwest National Laboratory, Richland, WA, USA.

S. Abedin, K., 2005. Observation of strong stimulated Brillouin scattering in single-mode As2Se3 chalcogenide fiber. Opt. Express 13, 10266–10271.

Santagiustina, M., Chin, S., Primerov, N., Ursini, L., Thévenaz, L., 2013. All-optical signal processing using dynamic Brillouin gratings. Sci. Rep. 3, 1594.

Shin, H., Qiu, W., Jarecki, R., Cox, J.A., Olsson, R.H., Starbuck, A., Wang, Z., Rakich, P.T., 2013. Tailorable stimulated Brillouin scattering in nanoscale silicon waveguides. Nat. Commun. 4, 1944.

Shin, H., Cox, J.A., Jarecki, R., Starbuck, A., Wang, Z., Rakich, P.T., 2015. Control of coherent information via on-chip photonic–phononic emitter–receivers. Nat. Commun. 6, 6427.

Solmaz, M., Park, H., Madsen, C.K., Cheng, X., 2008. Patterning chalcogenide glass by direct resist-free thermal nanoimprint. J. Vac. Sci. Technol. B 26, 606–610.

Song, K.Y., Hotate, K., 2007. 25 GHz bandwidth Brillouin slow light in optical fibers. Opt. Lett. 32, 217–219.

Spälter, S., Hwang, H.Y., Zimmermann, J., Lenz, G., Katsufuji, T., Cheong, S.W., Slusher, R.E., 2002. Strong self-phase modulation in planar chalcogenide glass waveguides. Opt. Lett. 27, 363–365.

Tan, W.C., Solmaz, M.E., Gardner, J., Atkins, R., Madsen, C., 2010. Optical characterization of a-As2S3 thin films prepared by magnetron sputtering. J. Appl. Phys. 107, 033524.

Van Laer, R., Kuyken, B., Van Thourhout, D., Baets, R., 2015. Interaction between light and highly confined hypersound in a silicon photonic nanowire. Nat. Photon. 9, 199–203.

Youden, K.E., Grevatt, T., Eason, R.W., Rutt, H.N., Deol, R.S., Wylangowski, G., 1993. Pulsed laser deposition of Ga-La-S chalcogenide glass thin film optical waveguides. Appl. Phys. Lett. 63, 1601–1603.

Zarifi, A., Stiller, B., Merklein, M., Liu, Y., Morrison, B., Casas-Bedoya, A., Ren, G., Nguyen, T.G., Vu, K., Choi, D.-Y., Mitchell, A., Madden, S.J., Eggleton, B.J., 2018. Brillouin spectroscopy of a hybrid silicon-chalcogenide waveguide with geometrical variations. Opt. Lett. 43, 3493–3496.

Zembutsu, S., Fukunishi, S., 1979. Waveguiding properties of (Se,S)-based chalcogenide glass films and some applications to optical waveguide devices. Appl. Optics 18, 393–399.

Zhu, Z., Dawes, A.M.C., Gauthier, D.J., Zhang, L., Willner, A.E., 2007. Broadband SBS slow light in an optical Fiber. J. Light. Technol. 25, 201–206.

CHAPTER SIX

Superfluid SBS

Warwick P. Bowen[a,*], Christopher G. Baker[a], and Jack G.E. Harris[b]
[a]Australian Research Council Centre of Excellence for Engineered Quantum Systems, School of Mathematics and Physics, The University of Queensland, St Lucia, QLD, Australia
[b]Department of Physics, Yale University, New Haven, CT, United States
*Corresponding author: e-mail address: w.bowen@uq.edu.au

Contents

1. Introduction

Since the prediction of Brillouin scattering 100 years ago (Brillouin, 1922; Mandelstam, 1926), recognition of the importance of Brillouin scattering, both fundamental and applied, has only grown. In the case where the Brillouin scattering medium is a classical solid or fluid,[a] the rich physics and

[a] That is, one that can be described by classical physics at the macroscopic level.

Semiconductors and Semimetals, Volume 109
ISSN 0080-8784
https://doi.org/10.1016/bs.semsem.2022.04.004

applications of Brillouin scattering are well described in the other chapters of this book, including existing applications in fiber optic sensing, material characterization, and biomedical science (chapters 4, 8, and 16), and emerging applications in areas such as silicon-chip-based sensors and quantum technologies (chapters 6, 11, and 13).

In this chapter, we provide an overview of Brillouin scattering from a quantum fluid, specifically superfluid helium. Superfluid helium exhibits quantum behaviors that are not present in regular fluids (Tilley and Tilley, 1990), including quantized circulation, a macroscopic wavefunction, quantum phase transitions, and quantum turbulence. The presence of a phase transition from normal to superfluid also introduces physics that is otherwise not accessible. Moreover, the vanishing viscosity of superfluid helium at low temperatures, combined with other extreme material properties, provides a set of characteristics that one may hope to leverage to build new technologies, particularly quantum technologies such as quantum interfaces, sensors, and memories.

Superfluid helium was discovered at around the same time as Brillouin scattering was predicted (London, 1938; Wolfke and Keesom, 1927). This led to a remarkable sequence of further discoveries and a continuous string of Nobel prizes into the present day. Brillouin scattering has played an important role in these discoveries, providing insight into the material properties of this rich quantum liquid. In recent times, the Brillouin interaction has begun to be exploited, not only to understand the properties of superfluid helium but also to control them via radiation pressure and other light forces (Harris et al., 2020; He et al., 2020; Kashkanova et al., 2017; Shkarin et al., 2019). In this chapter our aim is to encompass this full range of topics, from the historic contributions Brillouin scattering has made to our understanding of superfluid helium, to its new use to control superfluid dynamics and the future applications this may enable.

The early parts of the chapter are designed to provide an accessible introduction to superfluid helium and a historical account of early Brillouin scattering experiments. We then introduce recent progress on confining superfluid within or upon optical resonators and using the enhanced light-superfluid interactions this provides to control the behavior of sound waves within the fluid at both classical and quantum levels. Finally, we provide a perspective on how these enhanced Brillouin scattering interactions could be employed in future both to better understand the nature of superfluid helium and for quantum technological applications.

2. Basics of superfluid helium

For more than a century, superfluid helium has provided access to a remarkable range of physical phenomena, including Bose–Einstein condensation, phase transitions, quantized circulation, topological order, low dimensionality, and quantum turbulence, to name a few (Barenghi et al., 2001; Donnelly, 1991; Volovik, 2003). Access to this rich behavior is complemented by—and, to some extent, enabled by—superfluid helium's extreme material properties (Donnelly and Barenghi, 1998). In comparison with any other condensed material, superfluid helium offers the largest or smallest values of mass density, elastic modulus, surface tension, speed of sound, thermal conductivity, viscosity, anharmonicity, index of refraction, electronic bandgap, optical absorption, chemical reactivity, chemical purity, and isotopic purity. These extreme values are either inherent to superfluid helium (and hence readily accessible in any low-temperature laboratory), or may be attained through well-established means (as in the case of isotopic purity).

While superfluid helium can be used to access many complex phenomena, SBS in this material can be described in relatively straightforward terms. A helpful starting point for this description is to consider a macroscopic, simply connected sample of liquid ^{4}He in equilibrium at temperature $T = 0$. Such a system is a pure superfluid: a state of matter characterized by vanishing viscosity and a curlless velocity field. In equilibrium at $T = 0$, its density is uniform (i.e., $\rho(\mathbf{r}) = \rho_{\mathrm{He}}$, where $\rho_{\mathrm{He}} = 145\ \mathrm{kg/m^3}$ is the density of superfluid ^{4}He) and its velocity field is everywhere zero. The lowest energy excitations out of the superfluid state are density waves whose dispersion relation $\omega(\mathbf{k}) = v_{\mathrm{He}}|\mathbf{k}|$ in the limit $|\mathbf{k}| \rightarrow 0$ (here ω and $|\mathbf{k}|$ are the angular frequency and wavevector of a given density wave, and $v_{\mathrm{He}} = 238\ \mathrm{m/s}$ is the speed of sound in liquid ^{4}He). The existence of these modes may be understood as reflecting the fact that the Navier–Stokes equation (which governs the dynamics of a fluid) may be linearized for small velocities, displacements, and density variations to yield a scalar wave equation for $\rho(\mathbf{r})$ (Bohr and Mottelson, 1969; Schmidt, 2015). Equivalently, we can say that sound waves are among the many phenomena that a fluid can exhibit.

At this level of description, density waves in superfluid helium do not differ substantially from those in any other acoustically linear and homogeneous medium such as air or water. The superfluid state's vanishing viscosity does remove a source of damping that is present in other fluids, but a density wave in superfluid helium will nevertheless decay. This is because liquid helium's anharmonicity (the nonlinearity of the relation between pressure and density) adds nonlinear terms to the wave equation mentioned above. This causes an excitation with a given $\mathbf{k}$ to convert into excitations of other $\mathbf{k}'$ (Maris, 1977).

Each mode of these density waves (indexed by $\mathbf{k}$) represents (at least approximately) an independent harmonic oscillator. Thus, they may be quantized in essentially the same manner as the modes of the electromagnetic field (as described in standard texts on the subject). The quanta in each mode are known as *phonons*.

If we consider increasing the temperature slightly above zero, then these modes will be populated by phonons as described by the Bose–Einstein distribution. These phonons can be pictured as a gas of "particles" that move freely through a "vacuum" comprised of the pure superfluid. However, these "particles" are not strictly noninteracting: as mentioned above, the anharmonicity of liquid helium results in coupling between density waves. Thus, a given phonon will travel ballistically only for a finite distance before it scatters from one of the other phonons in the thermal population (and even at $T=0$, this same nonlinearity will eventually cause a phonon to spontaneously downconvert). Since the population of phonons capable of producing such scattering depends on T, so does the mean free path Λ of a phonon. Specifically, $\Lambda \propto T^{-4}$, where the coefficient of proportionality is such that $\Lambda \approx 1$ mm for $T \approx 470$ mK (Maris, 1977).

For a sample of size L, there will be a value of T above which $\Lambda < L$ and the phonon "gas" will become collisionally dense (in the sense that a given phonon is likely to scatter from another phonon before it scatters from the sample boundary). When T is increased enough that $\Lambda \ll L$, a given phonon will be scattered by other phonons so frequently that the phonon "gas" is essentially hydrodynamic.

This marks the boundary between two important regimes in the behavior of the excitations out of the superfluid: for $\Lambda > L$ a perturbation to the phonon distribution that is initially localized in space will spread out through the sample at the speed of sound. However for $\Lambda \ll L$, the phonons are so dense as to constitute a "fluid." Just as in familiar fluids (like air and water), this "phonon fluid" is comprised of elastically colliding "particles," and so possesses the features of a normal fluid, including viscosity and a velocity field that need not be curlless. This "normal fluid" is governed by a Navier–Stokes equation, which may in turn be linearized for small velocities, displacements, and density variations to yield a scalar wave equation for this fluid's density field. These waves are known as "second sound," and have a linear dispersion with a propagation speed that varies with temperature, but has a typical value ~20 m/s (Donnelly, 2009).

In practice, the thermal population of phonons begins to form an effectively hydrodynamic normal fluid for $T \approx 1$ K. Above this temperature, it may be viewed as a fluid that coexists with the superfluid. In this picture (known as the "two-fluid model"), liquid helium is described by a two-component Navier–Stokes equation, whose excitations include both the density waves ("first sound") and the waves in the phonon density ("second sound") described above. However, it is important to note that for T well below 1 K, the thermal population of phonons is not hydrodynamic, so there is no normal fluid or second sound—only a nearly ballistic gas of phonons.

Somewhat confusingly, the terms "third sound," "fourth sound," and "zero sound" do not refer to the logical extrapolation of this picture (i.e., to the harmonic modes in the effectively hydrodynamic gas of second sound excitations). Rather, third sound modes are specific to thin films of superfluid helium, fourth sound modes are specific to narrow channels (Atkins, 1959), and zero sound modes are specific to Fermi-degenerate ^{3}He.

The focus of this chapter is on the interaction of these modes (in particular, first, second, and third sounds) with optical waves. Our focus is on optical waves with frequencies low enough that any electronic excitations in the helium may be ignored. Since the lowest such excitation is ~19 eV (corresponding to light with wavelength $\lambda \sim 65$ nm), this comfortably covers the infrared, visible, and well into the ultraviolet.

In this regime, light interacts with the superfluid largely because the superfluid's excitations modulate its local index of refraction $n(\mathbf{r}, t)$. This modulation may reflect variations in the density (as for a first sound mode), or in the thickness of a film (as for a third sound mode). In the former case, a specific excitation $\delta\rho(\mathbf{r}, t)$ leads to $n(\mathbf{r}, t) = n_{\mathrm{He}}(1 + \delta\rho(\mathbf{r}, t)/\rho_{\mathrm{He}})$ where $n_{\mathrm{He}} = 1.028$ and $\rho_{\mathrm{He}} = 145$ kg/m^3 are the refractive index and density of superfluid ^{4}He in equilibrium at $T = 0$.

Thus, density and thickness fluctuations in superfluid helium alter the propagation of light. Correspondingly, variations in the light intensity exert pressure on the helium that can excite density and thickness modes (depending on the specific realization, this pressure is known as electrostriction, radiation pressure, or optical dipole force). These influences exerted by superfluid on light—and vice versa—are the focus of this chapter.

3. Early superfluid Brillouin experiments

3.1 Light scattered in superfluid helium

When Wolfke and Keesom (1927) observed that a phase transition occurred in cryogenically cooled helium-4, it was immediately recognized that light scattering experiments might provide important information about the properties of the different phases and about the transition between them. However, compared to other liquids, liquid helium only weakly scatters light, due both to its low temperature (and therefore low level of thermal excitations) and to its small polarizability. As a consequence, the first attempts to measure light scattering failed, with no signal observed over the noise floor of the measurement apparatus (McLennan et al., 1932).

Interest in light scattering was renewed when London suggested that the two-fluid behavior of superfluid helium-4 could be attributed qualitatively to Bose–Einstein condensation (London, 1938). This work focused on the light scattering properties of liquid helium-4 near the superfluid transition temperature (the λ point) (Findlay et al., 1940; Schiff, 1940). When fluids approach a second order or continuous phase transition, their isothermal compressibility diverges toward infinity. Small pressure fluctuations can then lead to wavelength-scale density fluctuations, which are referred to as critical fluctuations. This causes a large increase in light scattering termed *critical opalescence* (Gopal, 2000). This light scattering near the transition provides valuable information about the spectrum and intensity of critical fluctuations (Vinen, 1979), and as such was thought to provide a means to test London's proposal.

The first experiment to successfully observe light scattering from superfluid helium were performed by Jakovlev (1943), who employed a powerful mercury arc lamp and was able to observe the light scattering from superfluid helium-4 by eye. He found that the scattered light intensity was comparable to that from dry air, and that there was no significant change in intensity as the superfluid temperature was raised through the λ point. This latter observation was later confirmed via more quantitative light scattering measurements (Lawson and Meyer, 1954). The discovery of the laser enabled far more precise measurements (Grimm and Dransfeld, 1967). These experiments showed some change in light scattering across the λ point, but again no dramatic peak as would be expected from critical opalescence. The lack of observed critical opalescence was viewed as evidence against London's proposal that the superfluid helium phase transition was a form of Bose–Einstein condensation (Schiff, 1940). However, later measurements and calculations explained the lack of observed critical opalescence, showing that the divergence of the compressibility in helium-4 is weak, so that the temperature would need to be within 10 μK of the λ point to generate even a modest factor of two increase in light scattering (Vinen and Vaughan, 1970). To our knowledge, critical opalescence is yet to be observed near the λ point in pure helium-4. It has been observed on the gas-liquid phase transition in both helium-4 (Tominaga and Narahara, 1972) and helium-3 (Ohbayashi and Ikushima, 1973), and due to concentration fluctuations near the tricritical point in mixtures of helium-4 and helium-3 (Watts and Webb, 1974; Watts et al., 1972).

3.2 Spectrum of light scattering and the observation of Brillouin scattering

The previous section discussed measurements of the intensity of light scattered from superfluid helium. The spectrum of the scattered light also contains information about the fluid—information about the sound waves that can exist within it. The unique two-fluid behavior of superfluid helium results in the existence of a variety of sound waves that are not present in other fluids (discussed in Section 2), which can be explored and exploited using light scattering.

3.3 Brillouin scattering in superfluid helium-4

Brillouin scattering from thermally populated sound waves in superfluid helium was considered theoretically by Ginzburg in 1943 (Ginzburg, 1943). It was well known already at this time that Brillouin scattering could provide useful information about the properties of transparent media: specifically, allowing the velocity and attenuation of sound modes to be determined, and therefore the medium's viscoelasticity to be found, as discussed in regard to biological applications in chapter 16 of this book. Ginzburg, however, arrived at a striking conclusion. While scattering measurements on other media exhibit an elastic scattering peak at the frequency of the incident light due to Rayleigh scattering, together with frequency-shifted peaks due to Brillouin scattering, in superfluid helium-4 well below the λ point the Rayleigh scattering should be greatly suppressed, since no nonpropagating density fluctuations can exist. Indeed, the scattering spectrum from

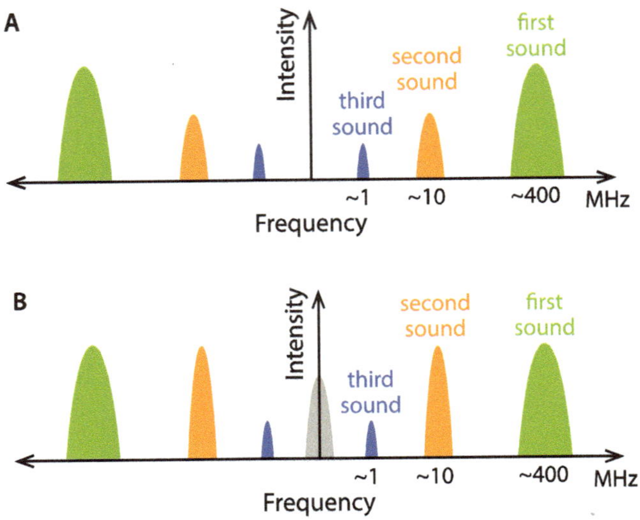

Fig. 1 Illustration of light scattering spectra in superfluid helium-4. (A) Pure helium-4; (B) a mixture of helium-4 and helium-3. The frequency axis corresponds to the frequency shift of the scattered light. *Green*: first sound; *orange*: second sound; *blue*: third sound; *gray*: Rayleigh scattering. The third sound peak should only exist in thin films of superfluid helium, and result from the van der Waals restoring force on the film due to the nearby presence of the substrate.

superfluid helium-4 should exhibit only peaks from first and second sound, and in thin films of superfluid from third sound also,[b] as illustrated in Fig. 1A.

The first observation of Brillouin scattering in superfluid helium was made by Woolf et al. (1966). They injected sound waves into superfluid helium-4 and measured the beat noise between light Brillouin scattered from first sound and the unscattered field, a beat note which occurs at the Brillouin frequency Ω. This was then be used to extract the velocity of sound v via the relation

$$\Omega/v = 2n\ k\ sin\theta \tag{1}$$

where n is the index of refraction, k is the optical wave number, and θ is the Bragg angle. Woolf et al. (1966)'s first sound velocity measurements were the first made in the hundreds of megahertz regime. Their results were consistent with lower frequency measurements in the 1–10 MHz range (Chase, 1958; Whitney and Chase, 1962), confirming that first sound has

[b] Brillouin peaks could also exist from capillary waves, called *ripplons*, on the surface of a deep pool of superfluid, where the restoring force results from surface tension.

no significant dispersion up to 723 MHz. They were also able to calculate the attenuation of the sound wave, which agreed reasonably well with expectations based on lower frequency experiments above the λ point (Abraham et al., 1966), and showed the same broad feature of a peak in attenuation expected from theory at the λ point (Khalatnikov and Chernikova, 1965, 1966).

The first observations of thermally driven Brillouin spectra in helium-4 were made by Pike et al. (1970) and St. Peters et al. (1970). As had been anticipated by Ginzburg (1943), the only Brillouin peaks that were observed were those due to first sound, with scattering from second sound expected have a far lower intensity. Both experiments also found that there was no visible Rayleigh peak at the frequency of the incident light, as also predicted by Ginzburg (1943). From their observed Brillouin spectra, Pike et al. (1970) were able to directly verify the results of Woolf et al. (1966), finding first sound velocity and attenuation consistent with this earlier work.

Later work was able to resolve both first and second sound peaks (O'Connor et al., 1975; Winterling et al., 1973). This allowed the dispersion and attenuation of second sound to both be determined at higher frequencies than had previously been possible. It also allowed the Landau–Placzek ratio to be determined. In regular fluids, this ratio is given by the ratio of the intensity of Rayleigh scattering due to thermally excited entropy fluctuations to the total intensity of the Brillouin doublet. It is important because it is equal to $C_P/C_V - 1$, where C_P and C_V are the heat capacity of the fluid at constant pressure and constant volume, respectively. Thus, the Landau–Placzek ratio provides important thermodynamical information about the properties of the fluid. In superfluid, since second sound can be associated to entropy fluctuations, the Landau–Placzek ratio is to good approximation equal to the ratio of the Brillouin scattering from first and second sound (Vinen, 1971).

3.4 Brillouin scattering in mixtures of helium-4 and helium-3

With the expectation raised by Ginzburg (1943) that it would be challenging to experimentally observe spontaneous Brillouin scattering from second sound, an alternative approach was investigated wherein a small concentration of helium-3 was mixed into the helium-4. Gor'kov and Pitaevskii (1958) and Ganguly and Griffin (1968) showed that in such mixtures, the relative concentrations of the two isotopes are strongly coupled to second sound. From this observation, they predicted that second sound in ^{3}He–^{4}He mixtures should exhibit strong Brillouin scattering. They also predicted that an unshifted peak should appear due to scattering from nonpropagating concentration fluctuations. Together, this would result in five Brillouin peaks, as shown in Fig. 1B, neglecting Brillouin scattering from surface waves such as third sound. These predictions were confirmed in a series of experiments (Palin et al., 1971a; Pike et al., 1969).

Pure ^3He superfluids have also been predicted to exhibit Brillouin scattering from first sound (Abrikosov and Khalatnikov, 1959). Here, the Fermionic nature of the fluid allows a further class of sound termed zero sound (Enss and Hunklinger, 2005; Keen et al., 1965), and therefore an additional pair of Brillouin peaks. However, the observation of Brillouin scattering in a helium-3 superfluid is made more challenging due to the much lower superfluid transition temperature. To date, to the authors knowledge, Brillouin scattering has yet to be observed in superfluid helium-3.

3.5 Stimulated Brillouin scattering

Stimulated Brillouin scattering offers an alternative approach to the measurements of scattering of coherent (Woolf et al., 1966) and thermal (Pike et al., 1970; St. Peters et al., 1970) excitations discussed earlier in this section of the chapter. Here, the incident optical field is strong enough to itself generate phonons within the superfluid. This generally occurs via electrostriction, but for surface waves can also arise due to fountain pressure (see He et al., 2020 and Section 5.2.3 for more details). These phonons then increase the levels of light scattered. Interference of the scattered light with the incident field reinforces the generation of phonons,cy the leading to a cycle of amplification. Once the gain exceeds the attenuation of the sound wave, a threshold occurs beyond which the sound wave oscillates much like a laser. A significant advantage of using stimulated rather than spontaneous Brillouin scattering is that the measurement speed can be greatly increased. However, this comes at the cost of requiring much higher optical intensities, intensities that can only practically be reached by using a pulsed laser (Djadaojee et al., 2021).

Stimulated Brillouin scattering was first observed in helium-4 by Winterling et al. (1968), shortly prior to the first observations via spontaneous Brillouin scattering from thermal excitations (Pike et al., 1970; St. Peters et al., 1970). From the Brillouin frequency, they were able to determine the sound velocity, and later they showed it was also possible to measure the attenuation (Heinicke et al., 1969). Stimulated Brillouin scattering from third sound waves has recently also been observed (He et al., 2020). This was achieved by confining a thin superfluid film on the surface of a micro-scale optical resonator, and thereby enhancing the light-sound interaction, and is discussed further in Section 5.

Very recently, Djadaojee et al. (2021) have developed a stimulated Brillouin gain spectroscopy approach to study superfluid helium. Here, two optical fields are injected, with a frequency difference that matches the frequency of the sound wave they wish to probe. The Brillouin interaction then acts to amplify the lower frequency field. By scanning the frequency difference, it is possible to map out the stimulated gain function, and determine both the frequency and attenuation of a given sound wave. Stimulated Brillouin gain spectroscopy has the advantage of both fast acquisition and low power requirements. Djadaojee and Grucker (2021) have used it to determine the equation of state of bulk superfluid helium-4.

4. Superfluid cavity optomechanics with bulk waves

4.1 Background

Interactions between the elastic modes of a body (whether solid or liquid) and the light passing through that body may take on many forms. The area of cavity optomechanics can be thought of as focusing on systems in which the optical modes and elastic modes are each confined to a cavity of some kind (Aspelmeyer et al., 2014; Bowen and Milburn, 2015). In this case, the spectra of elastic and optical modes are both discrete. In practice, cavity optomechanical systems are easiest to understand when it is possible to consider a single optical mode and its coupling to a single elastic mode.

The coupling between these modes is usually modeled by regarding the optical mode's resonance frequency as being tuned by the displacement of the elastic mode. As discussed extensively in chapters 2 and 3 as well as references (Aspelmeyer et al., 2014; Bowen and Milburn, 2015), this picture results in the well-known single-mode optomechanical coupling governed by the Hamiltonian $H_{\mathrm{OM}} = \hbar\omega_{\mathrm{opt}}(x)a^{\dagger}a + \hbar\omega_{\mathrm{el}}b^{\dagger}b$. Here, $\omega_{\mathrm{opt}}(x)$ is the oscillation frequency of the optical mode (with its dependence upon the elastic mode's displacement x written explicitly), ω_{el} is the elastic mode's oscillation frequency, and a and b are the annihilation operators for the optical and elastic mode, respectively. In writing H_{OM} this way (i.e., with the parameter ω_{opt} depending on the degree of freedom x), we have implicitly made a Born–Oppenheimer approximation; this approximation is typically very accurate for optical modes (with PHz frequencies) that are coupled to elastic modes (typically in the kHz–GHz range).

Typically one is interested in displacements x sufficiently small that $\omega_{\mathrm{opt}}(x)$ can be expanded to first order, so that $H_{\mathrm{OM}} = \hbar\omega_{\mathrm{opt},0} a^\dagger a + \hbar\omega_{\mathrm{el}} b^\dagger b + \hbar g_0 a^\dagger a (b^\dagger + b)$ where $\omega_{\mathrm{opt},0} = \omega_{\mathrm{opt}}(0)$ and the single-quantum optomechanical coupling rate $g_0 = (d\omega_{\mathrm{opt}}/dx)x_{\mathrm{ZP}}$. The zero-point motion of the elastic mode is $x_{\mathrm{ZP}} = \sqrt{\hbar/2m\omega_{\mathrm{el}}}$ where m is the effective mass of the elastic mode. A physical picture of H_{OM} may be had by imagining a simple Fabry–Perot cavity in which one of the end mirrors has mass m and is attached to a spring such that it oscillates at ω_{el}. However, the same H_{OM} arises whenever an elastic mode alters the optical length of an optical mode (for example, by altering the index of refraction between two fixed mirrors).

We note that H_{OM} can also be interpreted as the classical Hamiltonian function that governs the (nonquantized) optical and elastic modes if a and b are regarded simply as complex numbers giving the amplitudes of each mode. Indeed, the classical behavior of such a system offers access to a number of important effects. These include the all-optical tuning of the elastic mode (this is the well-known "optical spring" effect), and the use of these systems as sensitive transducers that convert signals between the mechanical and optical domains (Aspelmeyer et al., 2014; Bowen and Milburn, 2015).

In the quantum regime, the range of phenomena that can be accessed depends on ratio of the coupling rate g_0 to the system's loss rates. In most systems realized to date, g_0 is considerably smaller than the optical mode's damping (typically the linewidth of the optical cavity κ). To compensate for this, the optical cavity may be driven to a high-amplitude coherent state, in which case H_{OM} can be linearized as $H_{\mathrm{lin}} = \hbar\omega_{\mathrm{opt},0} a^\dagger a + \hbar\omega_{\mathrm{el}} b^\dagger b + \hbar g(a^\dagger + a)(b^\dagger + b)$ where now a is reinterpreted as the annihilation operator for fluctuations around a coherent state of amplitude $\alpha \in \mathbb{C}$, and the multiphoton coupling rate $g = |\alpha| g_0$ (see chapter 3 and Aspelmeyer et al., 2014 and Bowen and Milburn, 2015).

Systems description by H_{lin} has provided access to a range of quantum optical and quantum acoustic phenomena. For example, the elastic medium's response to the optical mode can be used to produce squeezed light (Brooks et al., 2012; Purdy et al., 2013b). The quantum backaction of the optical mode on the elastic mode can be used to reach quantum-limited readout of the elastic mode (Brahms et al., 2012; Purdy et al., 2013a;

Safavi-Naeini et al., 2012; Teufel et al., 2016) and to prepare it in a variety of quantum states with potential applications in sensing and in testing quantum mechanics in unprecedentedly massive objects (de Lépinay et al., 2021; Hong et al., 2017). However for the purposes of this chapter, we note that the mathematical forms of H_{OM} and H_{lin} are well known in the quantum optics literature, as are their potential applications. Here, we focus on the particular technical advantages that superfluid helium offers in realizing these goals.

4.2 Superfluid as an optomechanical material

As mentioned briefly above, the phenomena that can be accessed by a cavity optomechanical system are limited by a number of practical issues. Some of the most relevant ones include: optical absorption in the device (which limits κ and leads to heating of the elastic mode), damping of the elastic mode, difficulties in cooling the elastic mode (i.e., via a dilution refrigerator), achieving strong coupling between the optical and elastic modes, efficiently coupling light into the optical cavity in a cryogenic environment, and the tendency of an optical mode to couple to a large number of elastic modes, only some of which are engineered to have low damping.

To varying degrees, superfluid helium can be used to address all of these challenges. Simply considering them in order, we note that the large electronic band gap and the absence of chemical or structural defects in a typical sample of superfluid helium renders them essentially incapable of absorbing photons in the near-IR wavelength range used in most cavity optomechanical systems (Seidel et al., 2002). The elastic quality factor in first sound resonators has been shown to reach $Q > 10^8$ at low temperatures in carefully designed systems (De Lorenzo and Schwab, 2014, 2017). The thermal conductivity of liquid helium at low temperatures is unusually large (for an insulating material), and there are well-established techniques for equilibrating helium with cryogenic refrigerators (Pobell, 2007). The fact the helium remains a liquid at arbitrarily low temperatures (under its own vapor pressure) means that it can conformally fill (or coat) any optical resonator. In the particular case of a Fabry–Perot optical cavity operating in the paraxial limit, the mirrors that confine the optical mode will also confine acoustic modes in the superfluid (i.e., first sound) with essentially the same spatial profile. In fact, the families of optical modes and acoustic modes are sufficiently similar that the orthogonality of eigenfunctions ensures that a given optical mode couples only to a single acoustic mode (Kashkanova et al., 2017).

This serves to remove the "forest" of unwanted acoustic modes that are a source of complications in many cavity optomechanics experiments, while also maximizing the mode overlap (and hence g_0) between the optical mode and the acoustic mode to which it couples.

4.3 Superfluid optomechanics experiments

A number of groups have built devices that take advantage of these features to varying degrees (De Lorenzo and Schwab, 2014, 2017; Kashkanova et al., 2017; Shkarin et al., 2019; Souris et al., 2017; Spence et al., 2021). Here, we describe results from two very similar devices studied in our group (Kashkanova et al., 2017; Shkarin et al., 2019). Fig. 2 is a schematic of one of these devices. It consists of a Fabry–Perot cavity that is realized between the end faces of two optical fibers. The fiber ends that have been laser machined to each have a slight concavity and an ultrapolished surface (Hunger et al., 2012). A high-finesse dielectric mirror coating was then deposited on each mirror face. When the two fibers are carefully aligned

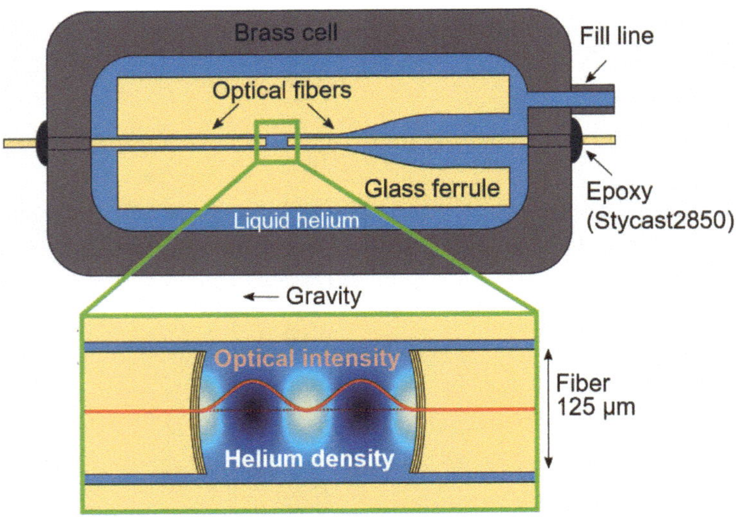

Fig. 2 Schematic illustration of a superfluid cavity optomechanical device. The optical cavity is formed between two optical fibers (*yellow*) that are aligned in a glass ferrule (*yellow*). The ferrule is mounted in a metal cell (*gray*). This cell is attached to a dilution refrigerator (not shown) and can be filled with superfluid helium (*blue*). The *lower panel* illustrates the optomechanical coupling: the intensity maxima of an optical mode (*red line*) overlap with the density maxima of an acoustic mode (*blue shading*, where darker corresponds to denser He). For clarity, this illustration shows a cavity with length $L = 2\lambda_{ac}$. In the actual device, $L = 200\lambda_{ac}$.

at a separation of not much more than 100 μm, they confine stable paraxial optical modes with wavelength (in the helium) $\lambda_{opt} \approx 1550$ nm. In addition, light traveling down the fibers is coupled into this mode with a mode-matching efficiency ~0.3 (Kashkanova et al., 2017; Shkarin et al., 2019). As a result, the optical cavity may be addressed via these fibers, which can be run out of the dilution refrigerator in a manner similar to conventional electrical wires. The removal of any need for *in situ* optical alignment is a considerable simplification over many other types of cryogenic optical cavities.

The optical cavity is housed in a superfluid leak-tight chamber that can be filled via a capillary line, which in turn is heat sunk to the mixing chamber of a dilution refrigerator using a silver-sintered heat exchanger. As described above, the mirrors that define the stable paraxial optical modes also define stable paraxial acoustic modes, and each such optical mode couples only to the acoustic mode with wavelength $\lambda_{ac} = \lambda_{opt}/2$ (Kashkanova et al., 2017; Shkarin et al., 2019).

Fig. 3 shows a measurement of a cavity resonance at the base temperature of the refrigerator, before (red) and after filling the cavity with superfluid helium. The presence of the superfluid shifts the cavity resonance by an

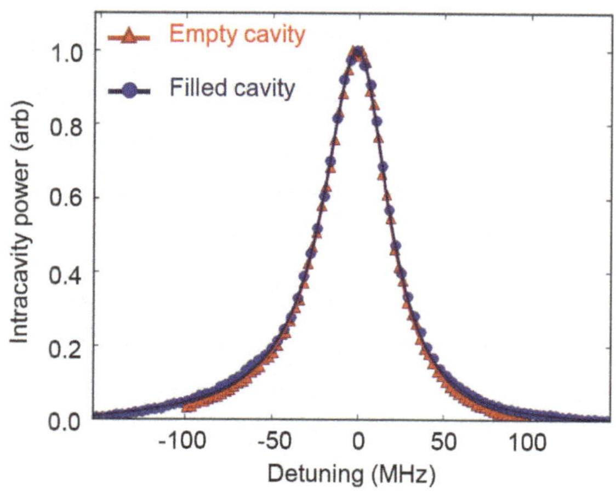

Fig. 3 Optical resonance of a superfluid-filled cavity. The data show measurements of the intracavity optical power as a function of the probing laser's detuning. *Red data*: the cavity contains no superfluid. *Blue*: the cavity has been filled with superfluid. In both cases, the cavity is cooled to the base temperature of the dilution fridge (~20 mK), To aid comparison between the empty and filled cavity, the two data sets are normalized to their maximum value and plotted as a function of the detuning from the cavity resonance. The solid *red* and *blue lines* are fits to the expected Lorentzian, and give linewidths 46.1 ± 0.1 MHz and 46.3 ± 0.2 MHz, respectively.

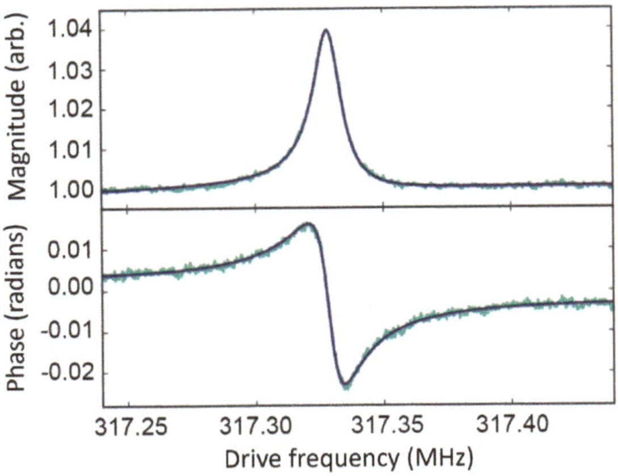

Fig. 4 The complex-valued response of a paraxial acoustic mode in a superfluid-filled cavity. The acoustic mode is driven by modulating the optical power in the cavity, and its response is inferred from the light leaving the cavity.

amount consistent with its index of refraction $n_{\mathrm{He}} = 1.028$ (this shift has been removed from the data), but does not change the linewidth of the mode. This is consistent with the very low value of optical absorption expected for liquid helium (Seidel et al., 2002).

Fig. 4 shows an acoustic resonance which is excited by and detected by intracavity light (via a process known as optomechanically induced transparency/amplification (Weis et al., 2010)). Fitting this data (as described in Kashkanova et al., 2017) provides the acoustic modes' resonance frequency and linewidth, as well as its optomechanical coupling rate g_0. The quality factor of this mode is shown as a function of the superfluid's temperature in Fig. 5. The black curve in this figure is a fit to the expected form, which consists of contributions from the anharmonicity of superfluid helium (mentioned in Section 2) and from the finite acoustic transmission of the mirrors.

Fig. 6 shows the Fourier spectrum of light that has been scattered from a similar device when no drive is applied to the acoustic mode. The red data show the Stokes light (corresponding to red-shifted light), while the blue data are the anti-Stokes light (corresponding to blue-shifted light). The green data in the upper panel are the real part of the cross-correlation between the Stokes and anti-Stokes signals, and the green data in the lower panel are the imaginary part of this cross-correlation. Independent measurements were used to normalize the data so that the height of the anti-Stokes feature corresponds to $\langle b^{\dagger} b \rangle$, the mean number of phonons in the acoustic mode.

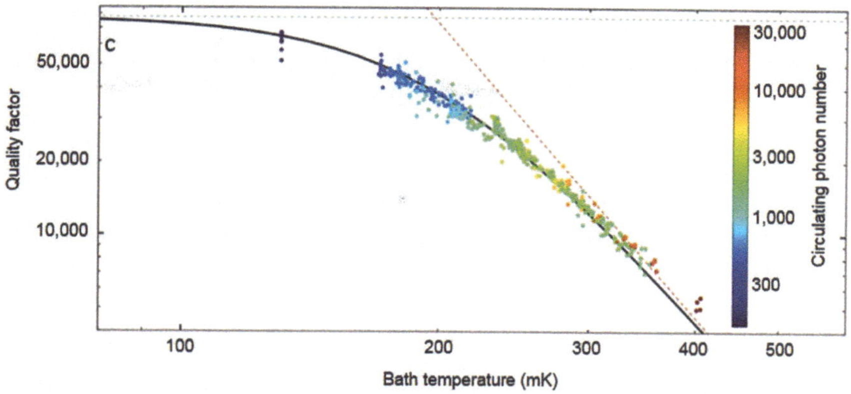

Fig. 5 The quality factor of the same acoustic mode as in Fig. 4, plotted as a function of the temperature T of the superfluid. As described in Kashkanova et al. (2017), T is controlled via the dilution refrigerator and the heat dissipated by the incident laser. The predicted contributions from three-phonon scattering and acoustic radiation into the fibers are shown as the *red* and *blue dashed lines*, respectively. Their combined effect is the black curve.

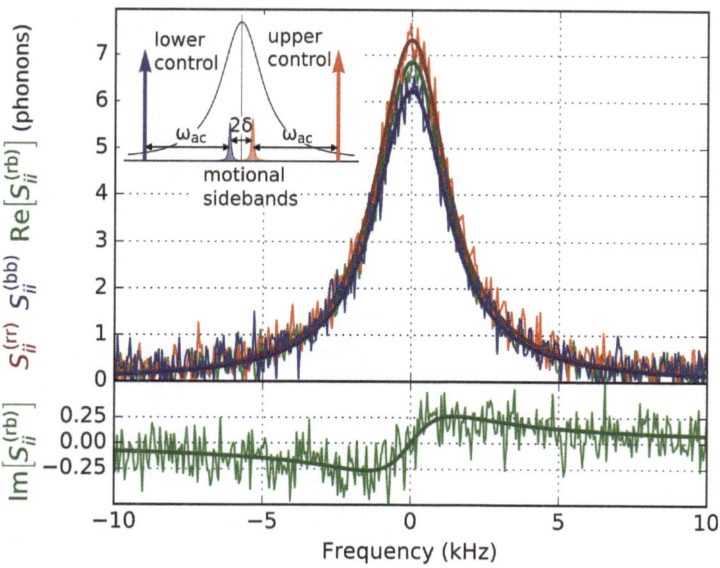

Fig. 6 Measurement of the thermal and quantum fluctuations of a single acoustic mode in a superfluid-filled cavity. In the inset, which illustrates the measurement scheme, the *black curve* is the cavity lineshape, the *red* and *blue arrows* are the probe lasers, and the *red* and *blue curves* are the Stokes and anti-Stokes sidebands that are measured via heterodyne scheme. The *upper panel* shows the spectrum of the Stokes and anti-Stokes sidebands (*red* and *blue*), along with the real part of their cross-correlation (*green*). The *lower panel* shows the imaginary part of their cross-correlation. The data are normalized and fit as described in the main text and in Shkarin et al. (2019).

The curves are a single fit of the entire data set to the form predicted by standard quantum optomechanics theory. The heights of the various features (specifically, that the three curves in the upper panel differ by 1/2, and that the curve in the lower panel has a peak-to-peak magnitude equal to 1/2) can be understood as arising from a combination of the acoustic mode's zero-point fluctuations and the quantum back-action of the readout provided by the optical mode (Børkje, 2016; Khalili et al., 2012; Weinstein et al., 2014).

5. Superfluid cavity optomechanics with surface waves

5.1 Background

Departing from the previous chapter, which considered bulk superfluid, we focus now on the coupling of light to acoustic waves existing in thin superfluid helium films. The basic experimental approach is as follows: an optical microcavity (or waveguide) is placed within a superfluid-tight enclosure held below the superfluid transition temperature. Addition of a small amount of helium leads to the formation of nanometer-thick self-assembling superfluid film, the thickness of which can be controlled by the volume of He gas injected. This film coats everything within the enclosure, including the microresonator itself. Indeed, the affinity of the superfluid for the substrate (through the van der Waals interaction) exceeds its affinity for itself, causing it to spread in order to maximize coverage. While this process is self-limiting for viscous liquids, such as liquid helium above the superfluid transition temperature, the absence of viscosity allows this to proceed here unimpeded, and accounts for the early observation of superfluid helium creeping out of vessels via nanometer-thick "Rollin films" (Rollin and Simon, 1939). Due to the absence of viscosity, superfluid acoustic waves which manifest as thickness (and temperature) fluctuations can exist even in films even as thin as a few monolayers (Scholtz et al., 1974). These are called third sound waves (Atkins, 1959) and resemble shallow water waves of wavelength λ in a fluid of depth $d \ll \lambda$. As in the case of water waves, their dispersionless velocity is $c \simeq \sqrt{gd}$, but in contrast to these the restoring force is provided not by gravity but by the van der Waals interaction with the substrate $g_{\mathrm{vdw}} = 3\alpha_{\mathrm{vdW}}/d^4$, where α_{vdW} is the van der Waals coefficient characterizing the interaction with the substrate.

Early superfluid third sound resonators were formed by confining superfluid within the plates of a capacitor. Motion of the superfluid between the

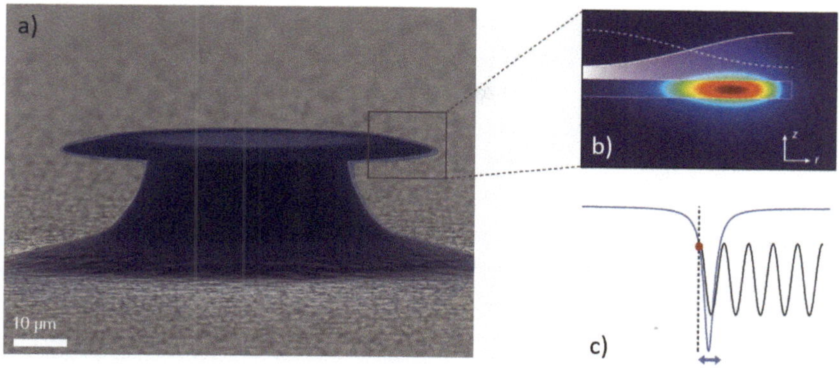

Fig. 7 Optomechanics with superfluid helium films. (A) Scanning electron microscope image of a silica microdisk on a silicon chip. The presence of a nm-thick superfluid He film in the experiments is materialized here by the blue transparent overlay. (B) Motion of the superfluid on the surface of the resonator perturbs the effective refractive index of the optical WGM, leading to dispersive optomechanical coupling encoding information about the superfluid motion into light leaving the cavity (C).

plates (dielectric constant $\varepsilon_r = 1.058$ (Donnelly and Barenghi, 1998)) modulated the capacitance, enabling both readout and electrostatic actuation of the film (Ellis and Hallock, 1983; Ellis and Li, 1993; Ellis and Luo, 1989; Schechter et al., 1998). More recently, optical detection and actuation schemes based upon the techniques of cavity optomechanics were introduced (Baker et al., 2016; Harris et al., 2016). The optomechanical interaction is illustrated in Fig. 7, where (A) shows a scanning electron microscope image of a silica microdisk whispering gallery mode (WGM) resonator on a silicon chip. The presence of a nm-thick superfluid He film in the experiments is materialized here by the blue transparent overlay. Since a fraction of the WGM optical field extends evanescently outside the cavity, motion of the superfluid on the surface of the resonator leads to a fluctuating amount of dielectric material in the optical mode's near field, perturbing its effective refractive index, as shown in (B). This leads to a dispersive optomechanical coupling, encoding information about the superfluid motion into the phase or amplitude of light leaving the cavity, as illustrated in Fig. 7C (Baker et al., 2016; Harris et al., 2016).

The approximately three orders of magnitude resonator size reduction compared to earlier capacitive resonators enabled by modern lithography and nanofabrication techniques, combined with the use of high-precision optical interferometric techniques, provided drastically improved sensitivity. These developments enabled the first real-time observation of superfluid

Brownian motion, as well as the first demonstration of laser cooling of a liquid (Harris et al., 2016), strong superfluid-mediated optical forces (McAuslan et al., 2016), and the first observation of the coherent dynamics of vortex ensembles in a superfluid helium film (Sachkou et al., 2019). More recently, we showed how optically induced surface deformations in a superfluid-coated microresonator provide strong Brillouin interactions, enabling an ultra-low Brillouin lasing threshold of 1.8 μW and strong phonon-mediated optical coupling (He et al., 2020; Pile, 2020). These results are discussed in the following.

5.2 Brillouin scattering in superfluid films

The general principle of Brillouin scattering with superfluid films is illustrated in Fig. 8A. Light confined within an optical waveguide (or resonator) interacts with a superfluid film coating the surface via its evanescent component, as discussed above. A superfluid acoustic wave, which manifests as a propagating thickness modulation, forms a traveling refractive index grating which can scatter the pump light. The beat between pump and scattered light creates regions of high mean optical intensity, which in turn draw in more superfluid through a combination of radiation pressure and fountain pressure interactions. This reinforces the original acoustic wave, in a dynamical feedback loop which can lead to lasing, as discussed in detail in this book's introductory chapters. The waveguide geometry shown in (a) can be wrapped around, forming a resonator. The self-assembling nature of the films lends itself naturally to a variety of resonator geometries, some of which are shown in Fig. 8B. In our work, we employed a silica microdisk, shown in Fig. 8B*ii*, where the same geometry which confines light also confines acoustic whispering gallery modes (WGMs) colocated with the optical WGMs close to the perimeter of the disk, as shown in Fig. 8C. In these thin film experiments, pump photons are scattered purely by moving boundary/radiation pressure-type interactions, and not by photoelastic/electrostrictive coupling (Baker et al., 2014), as superfluid helium flow in the third sound wave is considered incompressible. Indeed, while superfluid helium is in fact quite compressible, as discussed in Section 2, the van der Waals restoring pressure acting on the superfluid helium film's free surface is approximately three orders of magnitude lower than helium's bulk modulus (kPa range for the film thicknesses considered here (He et al., 2020) vs bulk modulus $K = 8$ MPa). As a consequence, any local influx of superfluid predominantly leads to a thickening of the film and not an increase in density at third sound

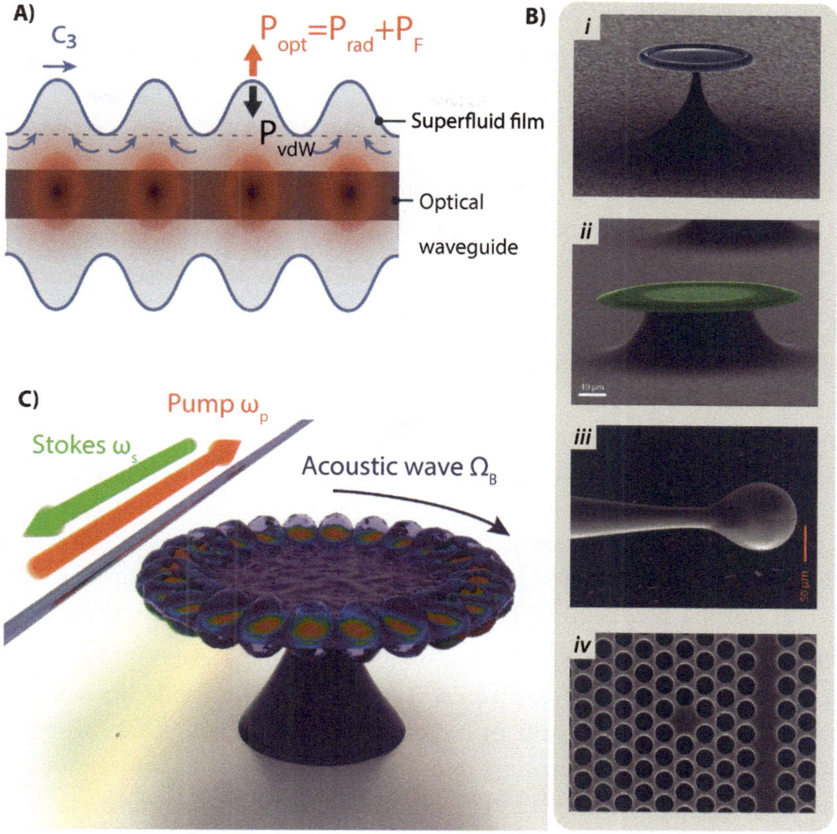

Fig. 8 Optomechanics and Brillouin scattering with superfluid helium films. (A) Interaction between an optical wave in a waveguide and a superfluid wave on its surface. Motion of the film is governed by the van der Waals interaction P_{vdW} and optical pressure P_{opt} composed of radiation pressure P_{rad} and fountain pressure P_F. (B) Examples of optical microresonators which can be functionalized with superfluid films. (*i*) Silica microtoroid (Harris et al., 2016; McAuslan et al., 2016; Sachkou et al., 2019); (*ii*) wedged silica microdisk (He et al., 2020); (*iii*) silica microsphere (Sawadsky et al., 2022); and (*iv*) silicon photonic crystal cavity, which may simultaneously also function as superfluid *phononic* crystal cavity (Sfendla et al., 2021). (C) Illustration of the backward Brillouin scattering process in the superfluid optomechanical resonator of radius $R = 40$ μm, where a pump photon of azimuthal order *m* (*red*) is scattered into a counterpropagating Stokes photon of order *m* (*green*) and a traveling acoustic wave phonon (*blue*) of azimuthal order $M = 2m$, where $m = kR$ is the azimuthal order of the WGM (He et al., 2020).

frequencies. While the fundamental characteristics of the Brillouin interaction are left unchanged in this superfluid implementation (i.e., a 3-wave mixing interaction between two photons of slightly different frequencies ω_p and ω_S and a mediating phonon of frequency Ω_B, where energy and momentum conservation dictate $\omega_p = \omega_S + \Omega_B$ and $k_p = k_S + k_B$), the (super)fluid nature of the film presents a number of key differences over its solid-state counterparts, which we will detail in the following.

5.2.1 Extreme compliance

First among these is the fluid interface's extreme compliance, compared to solid Brillouin materials. The compliance of fluid interfaces, leading to an enhanced response to optical forces, has been utilized in a number of applications, such as Brillouin scattering and optomechanics in liquid droplets (Dahan et al., 2016; Zhang and Chang, 1989). The superfluid nature of the film allows these desirable properties to be maintained down to micrometer- and nanometer-size scales, where viscosity becomes prohibitive in normal fluids. Here, the equivalent of the Young's modulus is the van der Waals pressure on the surface of the fluid, $P_{\text{vdW}} = \frac{3\rho\alpha_{\text{vdW}}}{d^3}$, which in these experiments is more than seven orders of magnitude smaller compared to common Brillouin materials (kPa range vs tens to hundreds of GPa)[c]. In addition to enhanced response to optical forces, this compliance leads to a lower Brillouin shift Ω in the MHz range, vs the gigahertz range in many solids at telecommunication wavelengths. Because of this small frequency shift, the Stokes beam is naturally resonant with the degenerate whispering gallery mode of same azimuthal order that counterpropagates with the pump mode. This removes the need for potentially complex spectral engineering of the resonator (Lee et al., 2012), and allows for triply resonant enhancement (pump, Stokes, and acoustic wave) which is independent of the device size, down to mode volumes in the range of the optical wavelength cubed (Harris et al., 2020). While such miniaturization is desirable as it enhances the strength of the Brillouin interaction, it has so far been hindered in conventional backward intramodal scattering by the inability to meet momentum and energy conservation in micrometer-scale devices whose free spectral range far exceeds the Brillouin shift.

[c] For an intuitive "feel" for this difference in compliance, one can relate the ease in deforming the surface of an inflated rubber balloon (pressure ~0.01 atm ~1 kPa) compared to the difficulty in straining a block of glass.

The radiation pressure P_{rad} which deforms the film outward is also referred to in the context of optical tweezers as an *optical gradient force*, which draws in dielectric material into regions of high electric field (Ashkin et al., 1986; Wiederhecker et al., 2009). For low index-contrast interfaces, such as the one between superfluid helium $\varepsilon_{\text{sf}} = 1.058$ and vacuum $\varepsilon = 1$, it takes the simple form $P_{\text{rad}} = -\frac{1}{2}\varepsilon_0(\varepsilon_{\text{sf}} - 1)E^2$, where E refers to the optical eigenmode electric field in the absence of superfluid. Remarkably, this can be made comparable in magnitude to P_{vdW} for modest optical powers, opening up the prospect of optically defined acoustic modes and waveguides, as well as the prospect of observing single-photon nonlinearities.

5.2.2 Optical strong coupling

In addition, above the Brillouin lasing threshold, the backscattering generated by the large amplitude Brillouin wave illustrated in Fig. 8C is sufficiently strong to enter a regime of mechanically mediated photon–photon strong coupling, that is, where photons cycle between the pump and Stokes resonances at a rate g faster than the decay rate of the optical fields, $g > \kappa, \Omega$. This phonon-mediated strong coupling manifests as a lasing-induced optical doublet splitting measured using a pump-probe scheme (He et al., 2020).

5.2.3 Fountain pressure

First observed by Allen and Jones (1938), and formalized by London (1939), the fountain effect is unique to superfluid systems. It arises due to heat-driven chemical potential differences which cause the superfluid to rush toward heat sources. Unlike the conservative radiation pressure force, this is an entropic force which arises as a consequence of the system's statistical tendency to evolve toward a state of maximal entropy, and leads to an excess pressure $P_F = \rho S \Delta T$ in response to a localized temperature increase ΔT, where S is the entropy per unit mass of the superfluid. Regions of high optical intensity, which draw in more superfluid due to the radiation pressure interaction, also tend to be hotter due to photon absorption. In our work (He et al., 2020), this resulted in an approximately one order of magnitude increase in the optical force exerted on the fluid film. To enhance this interaction further, the goal is—for a given thermal budget due to photon absorption—to simultaneously maximize the resulting pressure $P_F = \rho S \Delta T$, while at the same time operating in a regime where the thermal response time τ of the system is near the criterion $\Omega\tau \sim 1$ for optimal dynamical backaction (Metzger and Karrai, 2004). This is a complex optimization problem, as entropy S and thermal conductivities (and hence ΔT) vary by several orders of magnitude within the experimentally accessible temperature range (approximately 10 mK to 1 K). Recent work (Sawadsky et al., 2022) shows that fountain pressure's contribution to phonon gain can be made nearly eight orders of magnitude stronger than radiation pressure for lower frequency acoustic modes, leading to a record-low picowatt lasing threshold. Proper optimization might make this approach transferable to higher frequency MHz-range Brillouin waves.

5.2.4 Standing-wave vs traveling-wave Brillouin interaction

The fields of cavity optomechanics and Brillouin scattering have historically developed independently, adopting different terminologies to describe the inelastic scattering of light by acoustic waves. "Optomechanics" and "Brillouin scattering" have traditionally referred to standing-wave and traveling-wave interactions, respectively (see chapter 3 of this book for an in-depth discussion and attempt at unifying these two fields). In our system, shown in Fig. 8C, it is possible to transition between these two scenarios in situ, on the same device. This is achieved by varying the position of the coupling fiber taper, mediating a transition from a standing- to a traveling-wave interaction, enabled by a small amount of optical backscattering (He et al., 2020). This transition manifests in the optical power spectrum by the generation of symmetric and asymmetric optical sidebands, along with the generation of standing and traveling superfluid acoustic waves on the surface of the resonator, respectively (He et al., 2020).

5.3 Applications: Superfluid Brillouin vortex gyroscope

5.3.1 Background

Brillouin gyroscopes (Zarinetchi et al., 1991) have been shown to be excellent rotation sensors, with recent designs even reaching earth-rotation-rate sensitivity in a chip-scale implementation (Lai et al., 2020). These rely on the Sagnac phase shift between two counterpropagating optical beams and encode rotation information in the beat frequency between these two modes.

In a related fashion, superfluid Brillouin lasers offer the prospect of measuring not rotation but superfluid circulation, which is caused by quantized vortices, with unprecedented precision. This is an important goal since quantized vortices are central to the behavior of two-dimensional superfluids, as recognized by the 2016 Nobel Prize in Physics, but measuring their presence and dynamics is particularly challenging. Indeed, the normal-fluid core of a vortex in superfluid helium-4 is roughly 1 Å in size (Tilley and Tilley, 1990), the thickness of a superfluid helium film is typically less than

30 nm, and the refractive index of liquid helium is extremely close to that of vacuum ($n_{He} = 1.029$). Combined, these characteristics have precluded direct optical imaging. In bulk helium, imaging techniques have relied on the use of tracer particles which scatter light and are pulled in to the vortex core, such as micrometer-sized frozen hydrogen crystals (Bewley et al., 2006). Because of the nanometer thickness of 2D films, such an approach is impractical. Moreover, while direct measurement of vortex quantization has been made in bulk by Vinen (1961), similar measurements have yet to be made in superfluid helium films.

An alternate approach relies on measuring the vortices indirectly through their interaction with confined phonons. The interaction between a quantized vortex and a third sound wave is, in essence, an acoustic analog of the Sagnac effect in optics, whereby the frequencies of third sound waves shift depending on the net circulation in the film. A vortex on the surface of a microcavity introduces one quantum of circulation $\kappa = \frac{h}{m_{He}}$ to the superfluid, with m_{He} being the mass of a helium atom and h Planck's constant. A third sound wave traveling along the direction of circulation will be frequency upshifted, while a wave traveling against the vortex's flow field will be frequency downshifted (Ellis and Luo, 1989). Earlier work by our group employed this approach to allow the first imaging of coherent dynamics of vortex ensembles (Sachkou et al., 2019). It relied on simultaneously tracking the frequency splitting of multiple low azimuthal order third sound modes, whose broad spatial extent allowed for spatial imaging of the vortices' position, but hid their discrete nature (Forstner et al., 2019).

5.3.2 Proposed implementation

The concept of a Brillouin vortex gyroscope is schematically illustrated in Fig. 9A. The Brillouin shift $\Omega_B/2\pi = 2c/\lambda_{light}$ is proportional to the speed of sound c in the superfluid. The circulating flow field created by vortices alters this speed, enabling nondestructive readout of their presence. In a superfluid, the circulation $\oint \vec{v} \cdot \overrightarrow{dl} = n\kappa$ around a closed contour is quantized in terms of the circulation quantum $\kappa = h/m_{He}$, as shown in Fig. 9A (Tilley and Tilley, 1990). Importantly, this means the mean vortex flow velocity along the path of the Brillouin acoustic wave (dashed gray line) $\bar{v}_v = \frac{\kappa}{2\pi R}$ is independent of the vortex position within the confines of the resonator. This, unlike in our previous work (Sachkou et al., 2019), precludes imaging but provides a step-like frequency shift response upon the vortex radially traversing the acoustic mode, providing direct evidence for quantization.

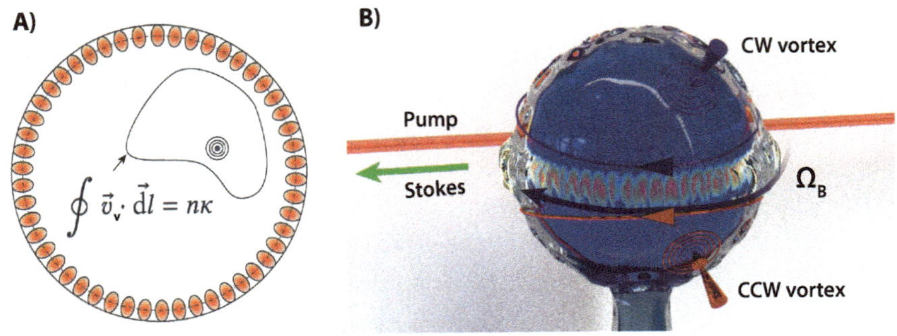

Fig. 9 Superfluid Brillouin gyroscope for vortex detection. (A) Circulation around any closed loop within the superfluid is quantized in terms of the quantum of circulation κ. The mean vortex flow velocity along the path of the Brillouin acoustic wave (*dashed gray line*) is therefore independent of the vortex position within the confines of the resonator. (B) Schematic of a practical implementation of a superfluid Brillouin gyroscope. Pump light generates an acoustic Brillouin wave of frequency Ω_B. Information on the superfluid circulation is encoded on the frequency of the Stokes wave $\omega_S = \omega_P - \Omega_B$. *Red* and *blue rings* show contours of constant circulation with CCW and CW vortices, respectively.

In a practical implementation, the resonator is typically not a 2D surface, but often takes a round three-dimensional shape, such as a microtoroid (Harris et al., 2016; McAuslan et al., 2016) or microdisk (He et al., 2020), conformally coated with superfluid. On a closed surface, any incompressible flow has zero total vorticity such that vortices are always present in pairs of opposite circulation (CW and CCW), as illustrated in Fig. 9B with a microsphere resonator. When both vortices are on the same hemisphere, their resulting flow fields cancel, resulting in no net change in the speed of sound c experienced by the Brillouin acoustic wave. However, when both vortices are on opposite sides of the equatorial plane, their contributions add up constructively, leading to a background flow with speed $2 \times \frac{\kappa}{2\pi R} \simeq 1$ mm/s for a 30 μm radius resonator. This results in a large discrete frequency shift $\Delta f = \frac{2\kappa}{\pi R \lambda_{\text{light}}} \sim 2\,\text{kHz}$ each time a vortex crosses the equatorial plane, well in excess of the ~1 Hz lasing acoustic linewidth (He et al., 2020).

6. Outlook

Superfluid helium is unique in many regards. It combines extreme physical properties (highlighted in Section 2) with a fluidic nature—rare among optomechanical and Brillouin systems—which enables it to conformally coat and fill arbitrary geometries and reservoirs, or self-assemble in the form of droplets which can be optically and magnetically levitated (Brown et al., 2021; Weilert et al., 1995). This enables its use in a wide latitude of experiments ranging from bulk to thin films and confined geometries. Its many acoustic degrees of freedom (first, second, third, and fourth sound) can couple to photons from the microwave (Lorenzo and Schwab, 2014) and visible (Palin et al., 1971b; St. Peters et al., 1970; Woolf et al., 1966) to the infrared (Harris et al., 2016; Kashkanova et al., 2017). For all these reasons, there is growing interest in its use for a wide range of future applications. To cite a few examples, superfluid helium resonators are being investigated as gravitational wave (Vadakkumbatt et al., 2021) as well as dark matter detectors (Carney et al., 2021). They have also been proposed as potential mechanical qubits (Sfendla et al., 2021) and as a means to study quantum turbulence, nonlinear hydrodynamics, quantized vortex dynamics (Sachkou et al., 2019), and solitons. Research is currently underway to further integrate superfluid functionality into superconducting (Koolstra et al., 2019) and photonic circuitry, while superfluid helium resonators' potentially low phonon occupancy (Patil et al., 2022) and frequency tunability make them a candidate of choice for use in future quantum networks.

Acknowledgments

W.P.B. and C.G.B. acknowledge funding from the US Army Research Office through grant number W911NF17-1-0310, and from the Australian Research Council Centre of Excellence for Engineered Quantum Systems (EQUS, project number CE170100009). C.G.B. acknowledges an Australian Research Council Fellowship DE190100318 and a UQ ECR Grant 2058737. J.G.E.H. acknowledges support by the NSF (Award No. 1707703), AFOSR (Grant No. FA9550-15-1-0270), the Vannevar Bush Faculty Fellowship (No. N00014-20-1-2628), the Office of Naval Research (MURI N00014-15-1-2761), and by the Quantum Information Science Enabled Discovery (QuantISED) for High Energy Physics (KA2401032).

References

Abraham, B.M., Eckstein, Y., Ketterson, J.B., Vignos, J.H., 1966. Measurements of the ultrasonic attenuation in liquid He 4 at low temperatures. Phys. Rev. Lett. 16 (23), 1039.

Abrikosov, A.A., Khalatnikov, I.M., 1959. The theory of a Fermi liquid (the properties of liquid 3He at low temperatures). Rep. Prog. Phys. 22 (1), 329.

Allen, J.F., Jones, H., 1938. New phenomena connected with heat flow in helium II. Nature 141 (3562), 243–244. http://www.nature.com/physics/looking-back/superfluid2/.

Ashkin, A., Dziedzic, J.M., Bjorkholm, J.E., Chu, S., 1986. Observation of a single-beam gradient force optical trap for dielectric particles. Opt. Lett. 11 (5), 288–290.

Aspelmeyer, M., Kippenberg, T.J., Marquardt, F., 2014. Cavity optomechanics. Rev. Mod. Phys. 86 (4), 1391.

Atkins, K.R., 1959. Third and fourth sound in liquid Helium II. Phys. Rev. 113 (4), 962–965. https://doi.org/10.1103/PhysRev.113.962.

Baker, C., Hease, W., Nguyen, D.-T., Andronico, A., Ducci, S., Leo, G., Favero, I., 2014. Photoelastic coupling in gallium arsenide optomechanical disk resonators. Opt. Express 22 (12), 14072–14086. https://doi.org/10.1364/OE.22.014072. http://www.opticsexpress.org/abstract.cfm?URI=oe-22-12-14072.

Baker, C.G., Harris, G.I., McAuslan, D.L., Sachkou, Y., He, X., Bowen, W.P., 2016. Theoretical framework for thin film superfluid optomechanics: towards the quantum regime. New J. Phys. 18 (12), 123025. https://doi.org/10.1088/1367-2630/aa520d. http://stacks.iop.org/1367-2630/18/i=12/a=123025.

Barenghi, C.F., Donnelly, R.J., Vinen, W.F., 2001. Quantized Vortex Dynamics and Superfluid Turbulence. Springer.

Bewley, G.P., Lathrop, D.P., Sreenivasan, K.R., 2006. Superfluid Helium: visualization of quantized vortices. Nature 441 (7093), 588. https://doi.org/10.1038/441588a. http://www.nature.com/nature/journal/v441/n7093/abs/441588a.html. 588.

Bohr, A., Mottelson, B.R., 1969. Nuclear Structure. W. A. Benjamin.

Børkje, K., 2016. Heterodyne photodetection measurements on cavity optomechanical systems: interpretation of sideband asymmetry and limits to a classical explanation. Phys. Rev. A 94 (4), 043816. https://doi.org/10.1103/PhysRevA.94.043816.

Bowen, W.P., Milburn, G.J., 2015. Quantum Optomechanics. CRC Press.

Brahms, N., Botter, T., Schreppler, S., Brooks, D.W.C., Stamper-Kurn, D.M., 2012. Optical detection of the quantization of collective atomic motion. Phys. Rev. Lett. 108, 133601. https://doi.org/10.1103/PhysRevLett.108.133601.

Brillouin, L., 1922. Diffusion de la lumière et des rayons X par un corps transparent homogène. Ann. Phys. 9 (17), 88–122.

Brooks, D.W.C., Botter, T., Schreppler, D.W.C., Purdy, T.P., Brahms, N., Stamper-Kurn, D.M., 2012. Strong optomechanical squeezing of light. Nature 488, 476–480. https://doi.org/10.1038/nature11325. https://rdcu.be/cGLNO.

Brown, C.D., Wang, Y., Namazi, M., Harris, G.I., Uysal, M.T., Harris, J.G.E., 2021. Characterization of levitated superfluid Helium drops in high vacuum. arXiv:2109.05618 [cond-mat, physics:physics]. http://arxiv.org/abs/2109.05618.

Carney, D., Krnjaic, G., Moore, D.C., Regal, C.A., Afek, G., Bhave, S., Brubaker, B., Corbitt, T., Cripe, J., Crisosto, N., Geraci, A., Ghosh, S., Harris, J.G.E., Hook, A., Kolb, E.W., Kunjummen, J., Lang, R.F., Li, T., Lin, T., Liu, Z., Lykken, J., Magrini, L., Manley, J., Matsumoto, N., Monte, A., Monteiro, F., Purdy, T., Riedel, C.J., Singh, R., Singh, S., Sinha, K., Taylor, J.M., Qin, J., Wilson, D.J., Zhao, Y., 2021. Mechanical quantum sensing in the search for dark matter. Quantum Sci. Technol. 6 (2), 024002. https://doi.org/10.1088/2058-9565/abcfcd.

Chase, C.E., 1958. Propagation of ordinary sound in liquid helium near the λ point. Phys. Fluids 1 (3), 193–200.

Dahan, R., Martin, L.L., Carmon, T., 2016. Droplet optomechanics. Optica 3 (2), 175. https://doi.org/10.1364/OPTICA.3.000175. https://www.osapublishing.org/abstract.cfm?URI=optica-3-2-175.

de Lépinay, L.M., Ockeloen-Korppi, C.F., Woolley, M.J., Sillanpää, M.A., 2021. Quantum mechanics-free subsystem with mechanical oscillators. Science 372 (6542), 625–629. https://doi.org/10.1126/science.abf5389.

De Lorenzo, L.A., Schwab, K.C., 2014. Superfluid optomechanics: coupling of a superfluid to a superconducting condensate. New J. Phys. 16 (11), 113020. https://doi.org/10.1088/1367-2630/16/11/113020.

De Lorenzo, L.A., Schwab, K.C., 2017. Ultra-high Q acoustic resonance in superfluid ^{4}He. J. Low Temp. Phys. 186 (3), 233–240. https://doi.org/10.1007/s10909-016-1674-x.

Djadaojee, L., Grucker, J., 2021. Optical measurement of the equation of state of bulk liquid helium-4 around 1 k. Phys. Rev. B 103 (14), 144513.
Djadaojee, L., Douillet, A., Grucker, J., 2021. Stimulated Brillouin gain spectroscopy of superfluid helium-4. J. Low Temp. Phys. 203 (1), 234–243.
Donnelly, R.J., 1991. Quantized Vortices in Helium II. Cambridge University Press.
Donnelly, R.J., 2009. The two-fluid theory and second sound in liquid helium. Phys. Today 62 (10), 34–39. https://doi.org/10.1063/1.3248499.
Donnelly, R.J., Barenghi, C.F., 1998. The observed properties of liquid helium at the saturated vapor pressure. J. Phys. Chem. Ref. Data 27 (6), 1217–1274. http://scitation.aip.org/content/aip/journal/jpcrd/27/6/10.1063/1.556028.
Ellis, F.M., Hallock, R.B., 1983. High resolution third sound resonator. Rev. Sci. Instrum. 54 (6), 751–753. https://doi.org/10.1063/1.1137468.
Ellis, F.M., Li, L., 1993. Quantum swirling of superfluid helium films. Phys. Rev. Lett. 71 (10), 1577. http://journals.aps.org/prl/abstract/10.1103/PhysRevLett.71.1577.
Ellis, F.M., Luo, H., 1989. Observation of the persistent-current splitting of a third-sound resonator. Phys. Rev. B 39 (4), 2703. http://journals.aps.org/prb/abstract/10.1103/PhysRevB.39.2703.
Enss, C., Hunklinger, S., 2005. Low-Temperature Physics. Springer, ISBN: 978-3-540-23164-6.
Findlay, J.C., Pitt, A., Smith, H.G., Goldstein, L., 1940. On Bose-Einstein fluids. Phys. Rev. 57 (3), 241.
Forstner, S., Sachkou, Y., Woolley, M., Harris, G.I., He, X., Bowen, W.P., Baker, C.G., 2019. Modelling of vorticity, sound and their interaction in two-dimensional superfluids. New J. Phys. 21 (5), 053029. https://doi.org/10.1088/1367-2630/ab1bb5.
Ganguly, B.N., Griffin, A., 1968. Scattering of light from entropy fluctuations in 3He-4He mixtures. Can. J. Phys. 46 (17), 1895–1903.
Ginzburg, V.L., 1943. Light scattering in helium II. Zh. Eksp. Teor. Fiz 13, 243.
Gopal, E.S.R., 2000. Critical opalescence. Resonance 5 (4), 37–45.
Gor'kov, L.P., Pitaevskii, L.P., 1958. On the scattering of light in He3-He4 mixtures. Sov. Phys. JETP 6 (33), 486–488.
Grimm, H., Dransfeld, K., 1967. The scattering of laser light in superfluid helium. Z. Naturforsch. A 22 (10), 1629–1630.
Harris, G.I., McAuslan, D.L., Sheridan, E., Sachkou, Y., Baker, C., Bowen, W.P., 2016. Laser cooling and control of excitations in superfluid helium. Nat. Phys. 12 (8), 788–793. https://doi.org/10.1038/nphys3714.
Harris, G.I., Sawadsky, A., Sfendla, Y.L., Wasserman, W.W., Bowen, W.P., Baker, C.G., 2020. Proposal for a quantum traveling Brillouin resonator. Opt. Express 28 (15), 22450–22461. https://doi.org/10.1364/OE.397478. https://www.osapublishing.org/oe/abstract.cfm?uri=oe-28-15-22450.
He, X., Harris, G.I., Baker, C.G., Sawadsky, A., Sfendla, Y.L., Sachkou, Y.P., Forstner, S., Bowen, W.P., 2020. Strong optical coupling through superfluid Brillouin lasing. Nat. Phys. 16 (4), 417–421. https://doi.org/10.1038/s41567-020-0785-0. https://www.nature.com/articles/s41567-020-0785-0.
Heinicke, W., Winterling, G., Dransfeld, K., 1969. Hypersonic absorption in liquid helium. Phys. Rev. Lett. 22 (5), 170.
Hong, S., Riedinger, R., Marinković, I., Wallucks, A., Hofer, S.G., Norte, R.A., Aspelmeyer, M., Gröblacher, S., 2017. Hanbury Brown and Twiss interferometry of single phonons from an optomechanical resonator. Science 358 (6360), 203–206. https://doi.org/10.1126/science.aan7939.
Hunger, D., Deutsch, C., Barbour, R.J., Warburton, R.J., Reichel, J., 2012. Laser microfabrication of concave, low-roughness features in silica. AIP Advances 2 (1), 012119. https://doi.org/10.1063/1.3679721.

Jakovlev, I.A., 1943. The scattering of light in liquid helium. J. Phys. USSR 7, 307.

Kashkanova, A.D., Shkarin, A.B., Brown, C.D., Flowers-Jacobs, N.E., Childress, L., Hoch, S.W., Hohmann, L., Ott, K., Reichel, J., Harris, J.G.E., 2017. Superfluid Brillouin optomechanics. Nat. Phys. 13 (1), 74–79.

Keen, B.E., Matthews, P.W., Wilks, J., Bleaney, B., 1965. The acoustic impedance of liquid helium-3. Proc. R. Soc. Lond. A Math. Phys. Sci. 284 (1396), 125–136. https://doi.org/10.1098/rspa.1965.0055.

Khalatnikov, I.M., Chernikova, D.M., 1965. RZh. Eksperim. Teor. Fix. 49, 1957.

Khalatnikov, I.M., Chernikova, D.M., 1966. RZh. Eksperim. Teor. Fix. 50, 411.

Khalili, F.Y., Miao, H., Yang, H., Safavi-Naeini, A.H., Painter, O., Chen, Y., 2012. Quantum back-action in measurements of zero-point mechanical oscillations. Phys. Rev. A 86, 033840. https://doi.org/10.1103/PhysRevA.86.033840.

Koolstra, G., Yang, G., Schuster, D.I., 2019. Coupling a single electron on superfluid helium to a superconducting resonator. Nat. Commun. 10 (1), 1–7. https://doi.org/10.1038/s41467-019-13335-7. https://www.nature.com/articles/s41467-019-13335-7.

Lai, Y.-H., Suh, M.-G., Lu, Y.-K., Shen, B., Yang, Q.-F., Wang, H., Li, J., Lee, S.H., Yang, K.Y., Vahala, K., 2020. Earth rotation measured by a chip-scale ring laser gyroscope. Nat. Photonics 14 (6), 345–349. https://www.nature.com/articles/s41566-020-0588-y.

Lawson, A.W., Meyer, L., 1954. Light scattering in liquid helium. Phys. Rev. 93 (2), 259.

Lee, H., Chen, T., Li, J., Yang, K.Y., Jeon, S., Painter, O., Vahala, K.J., 2012. Chemically etched ultrahigh-Q wedge-resonator on a silicon chip. Nat. Photonics 6 (6), 369–373. https://doi.org/10.1038/nphoton.2012.109. http://www.nature.com/nphoton/journal/v6/n6/abs/nphoton.2012.109.html.

London, F., 1938. The λ-phenomenon of liquid helium and the Bose-Einstein degeneracy. Nature 141 (3571), 643–644.

London, H., 1939. Thermodynamics of the thermomechanical effect of liquid He II. Proc. R. Soc. Lond. A Math. Phys. Sci. 171, 484–496. http://www.jstor.org/stable/97295.

Lorenzo, L.A.D., Schwab, K.C., 2014. Superfluid optomechanics: coupling of a superfluid to a superconducting condensate. New J. Phys. 16 (11), 113020. https://doi.org/10.1088/1367-2630/16/11/113020. http://iopscience.iop.org/1367-2630/16/11/113020.

Mandelstam, L.I., 1926. Light scattering by inhomogeneous media. Zh. Russ. Fiz-Khim. 58, 381–386.

Maris, H.J., 1977. Phonon-phonon interactions in liquid helium. Rev. Mod. Phys. 49, 341–359. https://doi.org/10.1103/RevModPhys.49.341.

McAuslan, D.L., Harris, G.I., Baker, C., Sachkou, Y., He, X., Sheridan, E., Bowen, W.P., 2016. Microphotonic forces from superfluid flow. Phys. Rev. X 6 (2), 021012. https://doi.org/10.1103/PhysRevX.6.021012.

McLennan, J.C., Smith, H.D., Wilhelm, J.O., 1932. XIV. The scattering of light by liquid helium. Lond. Edinb. Dublin Philos. Mag. J. Sci. 14 (89), 161–167.

Metzger, C.H., Karrai, K., 2004. Cavity cooling of a microlever. Nature 432 (7020), 1002–1005. http://www.nature.com/nature/journal/v432/n7020/abs/nature03118.html.

O'Connor, J.T., Palin, C.J., Vinen, W.F., 1975. Experimental determination of the Landau-Placzek ratio for liquid helium to the λ line. J. Phys. C Solid State Phys. 8 (2), 101.

Ohbayashi, K., Ikushima, A., 1973. Intensity of Rayleigh scattered light near the gas-liquid critical point of 3He. Phys. Lett. A 46 (2), 131–132.

Palin, C.J., Vinen, W.F., Pike, E.R., Vaughan, J.M., 1971a. Rayleigh and Brillouin scattering from superfluid 3He-4He mixtures. J. Phys. C Solid State Phys. 4 (10), L225.

Palin, C.J., Vinen, W.F., Pike, E.R., Vaughan, J.M., 1971b. Rayleigh and Brillouin scattering from superfluid 3He-4He mixtures. J. Phys. C Solid State Phys. 4 (10), L225. https://doi.org/10.1088/0022-3719/4/10/014. http://iopscience.iop.org/0022-3719/4/10/014.

Patil, Y.S.S., Yu, J., Frazier, S., Wang, Y., Johnson, K., Fox, J., Reichel, J., Harris, J.G.E., 2022. Measuring high-order phonon correlations in an optomechanical resonator. Phys. Rev. Lett. 128, 183601. https://journals.aps.org/prl/abstract/10.1103/PhysRevLett.128.183601.

Pike, E.R., Vaughan, J.M., Vinen, W.F., 1969. Brillouin scattering from first and second sound in a superfluid 3He-4He mixture. Phys. Lett. A 30 (7), 373–374.

Pike, E.R., Vaughan, J.M., Vinen, W.F., 1970. Brillouin scattering from superfluid 4He. J. Phys. C Solid State Phys. 3 (2), L40.

Pile, D., 2020. Superfluid coating. Nat. Photonics 14 (5), 267. https://doi.org/10.1038/s41566-020-0632-y. https://www.nature.com/articles/s41566-020-0632-y. 267.

Pobell, F., 2007. Matter and Methods at Low Temperatures, third ed. Springer, Berlin; New York, ISBN: 978-3-540-46356-6.

Purdy, T.P., Peterson, R.W., Regal, C.A., 2013a. Observation of radiation pressure shot noise on a macroscopic object. Science 339 (6121), 801–804. https://doi.org/10.1126/science.1231282.

Purdy, T.P., Yu, P.L., Peterson, R.W., Kampel, N.S., Regal, C.A., 2013b. Strong optomechanical squeezing of light. Phys. Rev. X 3, 031012. https://doi.org/10.1103/PhysRevX.3.031012.

Rollin, B.V., Simon, F., 1939. On the "film" phenomenon of liquid helium II. Physica 6 (2), 219–230. https://doi.org/10.1016/S0031-8914(39)80013-1.

Sachkou, Y.P., Baker, C.G., Harris, G.I., Stockdale, O.R., Forstner, S., Reeves, M.T., He, X., McAuslan, D.L., Bradley, A.S., Davis, M.J., Bowen, W.P., 2019. Coherent vortex dynamics in a strongly interacting superfluid on a silicon chip. Science 366 (6472), 1480–1485. https://doi.org/10.1126/science.aaw9229.

Safavi-Naeini, A.H., Chan, J., Hill, J.T., Alegre, T.P.M., Krause, A., Painter, O., 2012. Observation of quantum motion of a nanomechanical resonator. Phys. Rev. Lett. 108, 033602. https://doi.org/10.1103/PhysRevLett.108.033602.

Sawadsky, A., Harrison, R., Harris, G.I., Wasserman, W.W., Sfendla, Y.L., Bowen, W.P., Baker, C.G., 2022. Optimized entropic dynamical back action. (In preparation).

Schechter, A.M.R., Simmonds, R.W., Packard, R.E., Davis, J.C., 1998. Observation of 'third sound' in superfluid 3He. Nature 396 (6711), 554–557. https://doi.org/10.1038/25090. http://www.nature.com/nature/journal/v396/n6711/abs/396554a0.html.

Schiff, L.I., 1940. Scattering of light by liquid Helium. Phys. Rev. 57 (9), 844.

Schmidt, M., 2015. Optomechanical Arrays and Multimode Systems. Ph.D. thesis, University of Erlangen.

Scholtz, J.H., McLean, E.O., Rudnick, I., 1974. Third sound and the healing length of He II in films as thin as 2.1 atomic layers. Phys. Rev. Lett. 32 (4), 147–151. https://doi.org/10.1103/PhysRevLett.32.147.

Seidel, G.M., Lanou, R.E., Yao, W., 2002. Rayleigh scattering in rare-gas liquids. Nucl. Instrum. Methods Phys. Res. Sect. A 489 (1), 189–194. https://doi.org/10.1016/S0168-9002(02)00890-2.

Sfendla, Y.L., Baker, C.G., Harris, G.I., Tian, L., Harrison, R.A., Bowen, W.P., 2021. Extreme quantum nonlinearity in superfluid thin-film surface waves. npj Quantum Inf. 7 (1), 1–12. https://doi.org/10.1038/s41534-021-00393-3. https://www.nature.com/articles/s41534-021-00393-3.

Shkarin, A.B., Kashkanova, A.D., Brown, C.D., Garcia, S., Ott, K., Reichel, J., Harris, J.G.E., 2019. Quantum optomechanics in a liquid. Phys. Rev. Lett. 122 (15), 153601.

Souris, F., Rojas, X., Kim, P.H., Davis, J.P., 2017. Ultralow-dissipation superfluid micromechanical resonator. Phys. Rev. Appl. 7, 044008. https://doi.org/10.1103/PhysRevApplied.7.044008.

Spence, S., Koong, Z.X., Horsley, S.A.R., Rojas, X., 2021. Superfluid optomechanics with phononic nanostructures. Phys. Rev. Appl. 15, 034090. https://doi.org/10.1103/PhysRevApplied.15.034090.

St.Peters, R.L., Greytak, T.J., Benedek, G.B., 1970. Brillouin scattering measurements of the velocity and attenuation of high frequency sound waves in superfluid helium. Opt. Commun. 1 (9), 412–416. https://doi.org/10.1016/0030-4018(70)90166-5.

Teufel, J.D., Lecocq, F., Simmonds, R.W., 2016. Overwhelming thermomechanical motion with microwave radiation pressure shot noise. Phys. Rev. Lett. 116 (1), 013602. https://doi.org/10.1103/PhysRevLett.116.013602.

Tilley, D.R., Tilley, J., 1990. Superfluidity and Superconductivity. CRC Press, ISBN: 978-0-7503-0033-9.

Tominaga, A., Narahara, Y., 1972. Compressibility and correlation length of helium-4 near the critical point. Phys. Lett. A 41 (4), 352–354.

Vadakkumbatt, V., Hirschel, M., Manley, J., Clark, T.J., Singh, S., Davis, J.P., 2021. Prototype superfluid gravitational wave detector. Phys. Rev. D 104 (8), 082001. https://doi.org/10.1103/PhysRevD.104.082001.

Vinen, W.F., 1961. The detection of single quanta of circulation in liquid helium II. Proc. R. Soc. A Math. Phys. Eng. Sci. 260 (1301), 218–236. https://doi.org/10.1098/rspa.1961.0029.

Vinen, W.F., 1971. Light scattering by superfluid helium under pressure. J. Phys. C Solid State Phys. 4 (14), L287.

Vinen, W.F., 1979. Light scattering by liquid helium. Philos. Trans. R. Soc. Lond. A Math. Phys. Sci. 293 (1402), 377–390.

Vinen, W.F., Vaughan, J.M., 1970. The scattering of light from liquid helium. J. Phys. Colloq. 31 (C3), C3–29.

Volovik, G.E., 2003. The Universe in a Helium Droplet. Oxford Science Publications.

Watts, D.R., Webb, W.W., 1974. Critical opalescent light scattering in 3He-4He mixtures near the tricritical point. In: Low Temperature Physics-LT 13, Springer, pp. 581–585.

Watts, D.R., Goldburg, W.I., Jackel, L.D., Webb, W.W., 1972. Preliminary observations of light scattering from the 3He-4He mixture near its consolute critical point. J. Phys. Colloq. 33 (C1), C1–155.

Weilert, M.A., Whitaker, D.L., Maris, H.J., Seidel, G.M., 1995. Laser levitation of superfluid helium. J. Low Temp. Phys. 98 (1), 17–35. https://doi.org/10.1007/BF00754065.

Weinstein, A.J., Lei, C.U., Wollman, E.E., Suh, J., Metelmann, A., Clerk, A.A., Schwab, K.C., 2014. Observation and interpretation of motional sideband asymmetry in a quantum electromechanical device. Phys. Rev. X 4, 041003. https://doi.org/10.1103/PhysRevX.4.041003.

Weis, S., Rivière, R., Deléglise, S., Gavartin, E., Arcizet, O., Schliesser, A., Kippenberg, T.J., 2010. Optomechanically induced transparency. Science 330 (6010), 1520–1523. https://doi.org/10.1126/science.1195596.

Whitney, W.M., Chase, C.E., 1962. Velocity of sound in liquid helium at low temperatures. Phys. Rev. Lett. 9 (6), 243.

Wiederhecker, G.S., Chen, L., Gondarenko, A., Lipson, M., 2009. Controlling photonic structures using optical forces. Nature 462 (7273), 633–636. https://doi.org/10.1038/nature08584.

Winterling, G., Walda, G., Heinicke, W., 1968. Stimulated Brillouin scattering in liquid helium. Phys. Lett. A 26 (7), 301–302.

Winterling, G., Holmes, F.S., Greytak, T.J., 1973. Light scattering from first and second sound near the λ transition in liquid He. Phys. Rev. Lett. 30 (10), 427.

Wolfke, M., Keesom, W.H., 1927. Uber zwei verschiedene Flussigkeitszustande. Proc. Amst. 31, 81.

Woolf, M.A., Platzman, P.M., Cohen, M.G., 1966. Brillouin scattering in liquid helium II. Phys. Rev. Lett. 17 (6), 294.

Zarinetchi, F., Smith, S.P., Ezekiel, S., 1991. Stimulated Brillouin fiber-optic laser gyroscope. Opt. Lett. 16 (4), 229.

Zhang, J.-Z., Chang, R.K., 1989. Generation and suppression of stimulated Brillouin scattering in single liquid droplets. JOSA B 6 (2), 151–153. https://doi.org/10.1364/JOSAB.6.000151. https://www.osapublishing.org/josab/abstract.cfm?uri=josab-6-2-151.

Index

Note: Page numbers followed by "*f*" indicate figures and "*t*" indicate tables.

U

V

W

Z

FSC
www.fsc.org
MIX
Papier | Fördert
gute Waldnutzung
FSC® C083411

Kolibri 360 GmbH
Ferdinand-Jühlke-Straße 7
99095 Erfurt, Deutschland
produktsicherheit@kolibri360.de